Recycling of Waste Oils: Technology and Application

Recycling of Waste Oils: Technology and Application

Editors

Sebastiano Garroni
Alberto Mannu

MDPI • Basel • Beijing • Wuhan • Barcelona • Belgrade • Manchester • Tokyo • Cluj • Tianjin

Editors
Sebastiano Garroni
University of Sassari
Italy

Alberto Mannu
Mannu Consulting
Italy

Editorial Office
MDPI
St. Alban-Anlage 66
4052 Basel, Switzerland

This is a reprint of articles from the Special Issue published online in the open access journal *Processes* (ISSN 2227-9717) (available at: https://www.mdpi.com/journal/processes/special_issues/waste_oil).

For citation purposes, cite each article independently as indicated on the article page online and as indicated below:

LastName, A.A.; LastName, B.B.; LastName, C.C. Article Title. *Journal Name* **Year**, *Volume Number*, Page Range.

ISBN 978-3-0365-3595-1 (Hbk)
ISBN 978-3-0365-3596-8 (PDF)

Contents

About the Editors

Sebastiano Garroni is a research scientist with nearly 10 years of research experience in the synthesis and characterization of materials for energy storage and conversion, including piezoceramics for sensing. He obtained his Chemistry Degree (cum laude) on 2007 from the University of Sassari and he conducted his European PhD in Materials Science at the Universitat Autònoma de Barcelona, financed by an ITN Marie Curie Network Project titled COSY. After receiving his European PhD in 2011, he joined the physical chemistry group at the University of Sassari as a post doc fellow. In November 2012, he was contracted as an Assistant Professor in the Department of Chemistry and Pharmacy of the University of Sassari. He joined the ICCRAM-UBU research center (Burgos-Spain) in 2016 as a distinguished researcher within the European project Nanopiezoeletrics, financed by the Marie Sklodowska-Curie Individual Fellowship (IF) action (GA 707954). Since November 2021, he has been an Associate Professor (Physical Chemistry) at the University of Sassari. During these years, he gained a solid, broad and application-oriented background in the cross-fertilization of sustainable chemical synthesis and nanostructured materials for energy storage and conversion. His research track record is corroborated by 125 articles (H indx = 24) published in peer-reviewed journals (Scopus source) and an international patent (PCT 2013WO-IT00179), as well as attending international conferences in the fields of green energy vectors, mechanochemistry and functional materials, 20 times as an invited speaker and once to present a plenary.

Alberto Mannu is a chemist, who graduated from the University of Sassari (Italy) and then received a Ph.D. in homogeneous catalysis at the University of Barcelona (Spain). Through his academic career he has worked for top institutions, such as the Italian National Research Council (CNR), The Leibniz-Institut für Katalyse e. V. (Rostock, Germany), The Politecnico di Milano (Italy), and the University of Turin (Italy), before starting the professional firm Mannu Consulting in Milan. To the date, he is the author of about 50 papers in peer-reviewed journals, and he currently represents the advisory board for an international project funded under Horizon 2020 regarding the recycling of used vegetable oils. In 2022, he received an Italian National Scientific qualification as associate Professor in Chemistry for the disciplinary field of 03/B2—Principles of chemistry for applied technologies.

Editorial

Recycling of Waste Oils: Technology and Application

Alberto Mannu [1,2,*] and Sebastiano Garroni [2,*]

[1] Mannu Consulting, 20032 Milan, Italy
[2] Department of Chemistry and Pharmacy, University of Sassari, 07100 Sassari, Italy
* Correspondence: albertomannu@gmail.com (A.M.); sgarroni@uniss.it (S.G.)

Reducing the impact of human activity on the environment and, in general, on Earth, represents the most challenging target of the next years. It is mandatory to promote a sustainable development, able to cope from one side with the increasing demand of goods, and from the other with the availability of raw materials while preserving our ecosystems. In particular, increasing industrial production at any level is stressing the pool of available raw materials, turning many elements into critical.

Many efforts toward a sustainable development have been conducted by experts in different fields, including academic, industry, politics, and non-governmental organizations. As a general and transversal trend, the conversion of processes based on linear economy into circular one has been object of intensive work during the last years, and it represents one of the main targets of the next future. For implementing circular processes, it is mandatory to convert the term "waste" into "raw material". Nevertheless, this important change in common thinking should be accompanied by a specific progress in the available technologies which would allow to transform by-products into valuable goods. Furthermore, the conversion of these new raw materials in added-value products should be achieved in a sustainable way, both in terms of economics and environmental pollution.

In this context, a big family of potential raw materials it is represented by waste oils. These are typically generated from the lubricant chain (mineral waste oils), from the food chain (vegetable used oils), and from some industrial processes as the pyrolysis of plastics or from the extraction industry. The possibility to develop new and improved ways to recycle these products attracts the interest of many stakeholders from public and private sectors.

Citation: Mannu, A.; Garroni, S. Recycling of Waste Oils: Technology and Application. *Processes* **2021**, *9*, 2145. https://doi.org/10.3390/pr9122145

Received: 24 November 2021
Accepted: 24 November 2021
Published: 28 November 2021

In this Special Issue, a collection of recent studies about the recycling of oils are presented. The main topics here represented include:

- The waste generated from the prickly pear industry [1] and recovering and transformation of used vegetable oils (UVOs) [2].
- The exploitation of the oil produced from pyrolysis of plastic wastes [3].
- Some recent advances in polymer flooding technique for the enhanced oil recovery [4].

As the more representative topic is represented by the processing of UVOs, an overview about the technology currently available for its transformation [2], and the recently updated European Union legislation related to UVOs employment as raw materials [5], are also provided. The importance of this topic is due to the specific chemical composition of UVOs which, despite the waste nature of the product, is quite similar to the edible parent oil, basically composed by a mixture of glycerol-derived fatty acids. Fatty acids are known to be important raw materials as they can be easily converted into biodiesel, lubricants, and solvents. As the specific chemical composition of the collected UVOs can influence their destination, fast and low-invasive techniques of analysis are needed. In this regard, the application of several Nuclear Magnetic Resonance (NMR) techniques to the characterization of the fatty acid profile of UVOs is of high interest, as described by Di Pietro and coworkers in their review paper [6].

Apart from NMR-based techniques, which are able to furnish relevant information about the chemical composition of an oil, important insights about the properties of the minor components present in the raw material can be achieved by the analysis of the volatile fraction, as well as by the determination of the in vitro antioxidant activity or by the analysis of the total phenolic content [1].

In the end, the challenge to turn into sustainable and low impacting many processes which involve the employment of waste oils as raw materials is still open, and the progresses reached to date represents the basis for the milestones of the future.

Author Contributions: Writing—original draft preparation, A.M.; writing—review and editing, A.M.; S.G. All authors have read and agreed to the published version of the manuscript.

Funding: This research received no external funding.

Institutional Review Board Statement: Not applicable.

Informed Consent Statement: Not applicable.

Data Availability Statement: Not applicable.

Conflicts of Interest: The authors declare no conflict of interest.

References

1. Karabagias, V.K.; Karabagias, I.K.; Gatzias, I.; Badeka, A.V. Prickly pear seed oil by shelf-grown cactus fruits: Waste or maste? *Processes* **2020**, *8*, 132. [CrossRef]
2. Mannu, A.; Garroni, S.; Ibanez Porras, J.; Mele, A. Available technologies and materials for waste cooking oil recycling. *Processes* **2020**, *8*, 366. [CrossRef]
3. Kim, S.J. Comparison of Dimethylformamide with Dimethylsulfoxide for Quality Improvement of Distillate Recovered from Waste Plastic Pyrolysis Oil. *Processes* **2020**, *8*, 1024. [CrossRef]
4. Wang, X.; Xu, X.; Dang, W.; Tang, Z.; Hu, C.; Wei, B. Interception characteristics and pollution mechanism of the filter medium in polymer-flooding produced water filtration process. *Processes* **2019**, *7*, 927. [CrossRef]
5. Ibanez, J.; Martel Martín, S.; Baldino, S.; Prandi, C.; Mannu, A. European Union legislation overview about used vegetable oils recycling: The spanish and italian case studies. *Processes* **2020**, *8*, 798. [CrossRef]
6. Di Pietro, M.E.; Mannu, A.; Mele, A. NMR determination of free fatty acids in vegetable oils. *Processes* **2020**, *8*, 410. [CrossRef]

Communication

Comparison of Dimethylformamide with Dimethylsulfoxide for Quality Improvement of Distillate Recovered from Waste Plastic Pyrolysis Oil

Su Jin Kim

Department of Chemical Engineering, Chungwoon University, Incheon 22100, Korea; sujkim@chungwoon.ac.kr

Received: 4 April 2020; Accepted: 19 August 2020; Published: 21 August 2020

Abstract: As a part of improving the quality of the distillate (distilling temperature 120–350 °C) recovered from waste plastic pyrolysis oil (WPPO) by simple distillation, the enrichment of paraffin components present in the distillate was compared by the equilibrium extraction of dimethylformamide (DMF) and dimethylsulfoxide (DMSO). Regardless of the solvent used, the concentration increase rate of the paraffin component in the raffinate relative to the raw material was reduced by increasing the mass fraction of water in the solvent in an initial state. On the other hand, it increased by increasing the mass ratio of the solvent to the raw material in an initial state. The enrichment performance of paraffin component in raffinate recovered by DMF was higher than that by DMSO under the same experimental conditions. Furthermore, the two solvents were compared by adding color and the waxing phenomena of recovered raffinate to assess the enrichment performance of paraffin components.

Keywords: waste plastic pyrolysis oil; paraffin enrichment; DMF equilibrium extraction; DMSO equilibrium extraction

1. Introduction

The effective treatment of plastic waste products is extremely valuable in that it can solve the problems of serious environmental pollution and resource depletion occurring all over the world. Currently, the most common method of treating waste plastics is landfilling and incineration, but it is recognized that treatment through landfill and incineration is not the ultimate method for treating waste plastics because it causes various environmental problems [1]. Therefore, in recent years, research on the high-temperature thermal decomposition method as one of the efficient treatment of waste plastic products has been actively conducted from the point of view that it is eco-friendly and, at the same time, can be used in the recycling of resources [2].

A large amount of paraffin, olefin, aromatic and acid components are contained in mixed waste plastic pyrolysis oil (WPPO) such as polyethylene (PE), polypropylene (PP), polyvinyl chloride (PVC) and polyethylene terephthalate (PET). Until now, many studies, such as reaction by pyrolysis [2–8] only and pyrolysis reactions [1,9–13] using a catalyst to produce high quality WPPO for use as fuel, have been reported. Moreover, many studies have been reported that examine the possibility of using WPPO as a renewable energy by comparing with commercial fuels [14–16], but few studies have been conducted to improve the quality of the distillate recovered through a distillation operation for low-grade WPPO.

To investigate the quality improvement of the distillate recovered from WPPO for the purpose of producing products similar to commercial light oil, as shown in Figure 1a, it is necessary to examine the separation between paraffin components C_7–C_{24} contained in the distillate and other components using methods such as solvent extraction [17–23] and emulsion liquid [24–27], which are widely used for the dearomatizing of petroleum and coal tar systems.

This study investigates the quality improvement of distillate (distilling temperature 120–350 °C) recovered by the simple distillation of WPPO produced from four kinds of mixed waste plastics (PE, PP, PVC, PET). Using an aqueous solution of dimethylformamide (DMF) and dimethylsulfoxide (DMSO) as a solvent, the effects of liquid–liquid contacting time (t), the mass fraction of water in solvent of an initial state ($y_{w,0}$) and the mass ratio of solvent to the raw material of an initial state $(S/F)_0$ on the enrichment of paraffin components contained in the raw material (WPPO distillate) were examined. The comparison of the two solvents took into account the ease of phase separation, the enrichment performance of the paraffin components, their color, and the waxing phenomena of the recovered raffinate. Based on these results, I examined whether the final product (solvent-free raffinate) recovered in this study could be applied as commercial light oil.

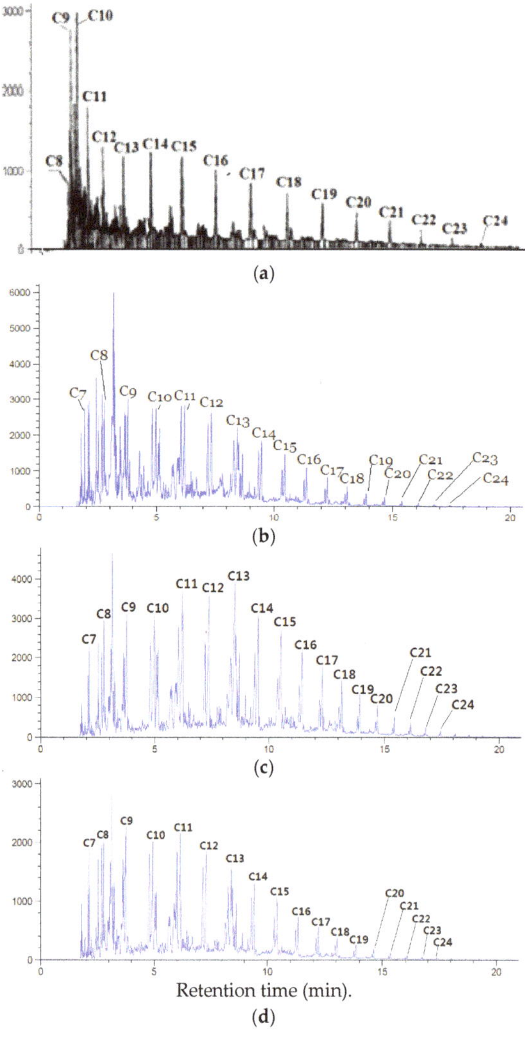

Figure 1. Gas chromatogram of (**a**) commercial light oil, (**b**) raw material (distillate recovered from WPPO through simple distillation), (**c**) raffinate (solvent-free) recovered using dimethylformamide (DMF), and (**d**) raffinate (solvent-free) recovered using dimethylsulfoxide (DMSO). Experimental conditions of (**c**) and (**d**): t = 72 h, $y_{w,0} = 0.03$, $(S/F)_0 = 10$, $T_0 = 28$ °C.

2. Material and Methods

2.1. Material

WPPO (pyrolysis raw material: mixed waste plastic of PE, PP, PVC and PET, reactor: batch pyrolysis furnace, reaction temperature: 430–550 °C) was supplied by Omega Energy Environmental Technology Co., Ltd. (Incheon, Korea). DMF (Anhydrous, ≥99.8% purity), paraffin component C_8 (octane, ≥99.7% purity), C_9 (*n*-nonane, ≥99% purity), C_{10} (*n*-decane, ≥99.8% purity), C_{12} (*n*-dodecane, ≥99.8% purity), C_{14} (*n*-tetradecane, ≥99.5% purity) and C_{16} (*n*-hexadecane, ≥99.8% purity) were purchased from Sigma-Aldrich Korea (Seoul, Korea).

2.2. Method

A certain concentration of solvent was prepared by the addition of distilled water. A 500 mL Erlenmeyer flask containing a certain amount of the raw material and solvent was placed in a shaking water bath maintained at an experimental temperature and vibrated for 12–72 h to reach liquid–liquid equilibrium. After the equilibrium was reached, the vibration was stopped and settled. The raffinate phase and the extract phase were separated using a 500 mL separatory funnel, and the mass of each phase was measured. The two phases were analyzed by adding acetone, and their compositions were determined. An analysis of two phases recovered was performed using a gas chromatograph (GC, Agilent Technologies Korea Co. (Seoul, Korea), 6890N) equipped with a flame ionization detector (FID) and column (J & W Co., DB–1HT (length: 30 m, inner diameter: 0.32 mm, film: 0.1 μm)). The analysis conditions of samples were as follows: carrier gas, N_2; flow rate, 2.2 mL·min^{-1}; injection port temperature, 300 °C; sample size, 2 μL; splitting ratio, 20:1; initial temperature, 40 °C; retention time, 1 min; rising temperature speed, 12 °C·min^{-1} and final temperature, 350 °C.

2.3. Material System and Conditions

Table 1 shows the material system and experimental conditions used in this study. Low-grade distillate (distilling temperature: 120–350 °C), which recovered by simple distillation of WPPO supplied from the company mentioned above, was used as the raw material. Aqueous solutions of DMF and DMSO were used as the solvents. The operating temperature of an initial state (T_0) was fixed constantly and t, $y_{w,0}$ and $(S/F)_0$ were changed. In this study, commercial reagent grade DMF, DMSO, *n*-octane, *n*-nonane, *n*-decane, *n*-dodecane, *n*-tetradecane, and *n*-hexadecane were used without further purification.

Table 1. Material system and experimental conditions.

System	
Raw material	Distillate [a] of Waste Plastic Pyrolysis Oil (WPPO)
Solvent	(1) aqueous solution of dimethylformamide (DMF) (2) aqueous solution of dimethylsulfoxide (DMSO)
Experimental Conditions	
Liquid-liquid contacting time, t (h)	12–72
Mass fraction of water in solvent of initial state, $y_{w,0}$ (−)	0–0.06
Mass ratio of solvent to raw material of initial state, $(S/F)_0$ (−)	1–10
Operating temperature of initial state, T_0 (°C)	28

[a] disatillate (distilling temperature: 120–350 °C) recovered by simple distillation of waste plastic pyrolysis oil (WPPO).

3. Results and Discussion

3.1. Gas Chromatogram of Raw Material (WPPO Distillate)

The raw material, which is composed of more than 100 components, contains paraffin, olefin, aromatic and acid components. Figure 1b shows a gas chromatogram of the raw material analyzed using capillary column DB–1HT and the carbon number of the 18 kinds of paraffin components (C_7–C_{24}) identified. The above-mentioned 18 kinds of paraffin components corresponded with peak patterns shown in the gas chromatogram, the result of GC mass analysis of the raw material and result analyzed by adding a small amount of the standard reagents of 6 kinds of paraffin components (C_8, C_9, C_{10}, C_{12}, C_{14}, C_{16}) purchased from Sigma-Aldrich Korea (Seoul, Korea).

The concentration of C_7–C_{20} paraffin components present in the raw material were in the range of about 0.04–2.77%, except for the concentration for the paraffin components of C_{21}–C_{24}, which were very small.

In this study, it was difficult to accurately quantify each paraffin component contained in the raw material, so the area % of the peak in the gas chromatogram was treated as the concentration of each paraffin component [28]. As shown in Figure 1a, the paraffin components (C_7–C_{24}) contained in the raw material of this study were almost identical to those contained in the commercial light oil (C_8–C_{24}).

3.2. Equilibrium Extraction

To confirm the time to reach equilibrium, the raffinate phases and the extract phases obtained through contacting (liquid–liquid contacting time: 12, 24, 48, 72 h) the solvent and the raw material under the same conditions [$y_{w,0}$ = 0.03, $(S/F)_0$ = 3, T_0 = 28 °C] were analyzed. The compositions of the raffinate phases and the extract phases obtained by contact for 48 and 72 h were identical to one another regardless of the contact time. The equilibrium arrival time was found to be less than 48 h. Therefore, equilibrium extraction was performed under the fixed condition of t = 72 h.

Figure 2a–d, respectively, show the effect of $y_{w,0}$ on the recovery rate of paraffin component i ($Y_{i,P}$, i = C_{12}, C_{16}), the recovery rate of the raffinate (Y_R), the concentration increase rate of the paraffin component i ($x_{i,P}$) and the densities (ρ) of the raffinate and extract phase when using DMF and DMSO. $Y_{i,P}$, Y_R and $x_{i,P}$, respectively, are defined as:

$$Y_{i,P} = \frac{R \times x_i}{R_0 \times x_{i,0}} \times 100\% \tag{1}$$

$$Y_R = \frac{R}{R_0} \times 100\% \tag{2}$$

$$x_{i,P} = \frac{x_i}{x_{i,0}} \times 100\% \tag{3}$$

where R and R_0, respectively, represent the mass of the extract phase after the liquid–liquid contact run and that of the raw material. $x_{i,0}$ and x_i denote the mass fraction of the paraffin component i in the raw material and that in raffinate recovered after a liquid–liquid contact run, respectively.

Regardless of the solvent used, an increase in $y_{w,0}$ was found to increase $Y_{i,P}$, as shown in Figure 2a. It is believed that this is the result of the increase in the extract phase polarity as $y_{w,0}$ increases. At the same $y_{w,0}$, the larger the carbon number of the paraffin component, the lower the solubility in the solvent, so the $Y_{i,P}$ increases. In addition, $Y_{i,P}$ in DMSO extraction showed a much higher value than that of DMF extraction in the entire range of $y_{w,0}$ in this study. The results showed that the DMF extraction method had better a solubility in paraffin components than the DMSO extraction method. In the case of DMF at $y_{w,0}$ = 0, the $Y_{i,P}$ values of paraffin components C_{12} and C_{16} were about 42.7% and 60.5%, respectively, whereas DMSO showed about 79% and 93.2%, respectively. In the case of DMSO, considering that the $Y_{i,P}$ of the C_{16} paraffin component is 97.8% at $y_{w,0}$ = 0.06, it is found that the paraffin components greater than the C_{16} paraffin component contained in the raw material remain at almost 100% in the raffinate recovered at $y_{w,0}$ = 0.06.

Figure 2b shows that Y_R increases as $y_{w,0}$ increases, regardless of the solvent used. It is considered that this is caused by decreasing the solubility of all the components such as paraffin, aromatic components, etc., contained in the raw material. In the entire range of $y_{w,0}$ in this study, DMSO extraction showed a much higher Y_R than DMF extraction. From this, it is considered that DMF extraction is superior to DMSO extraction in the solubility of paraffin components. For DMF, the Y_R values at $y_{w,0}$ = 0 and 0.06 were approximately 38% and 72.2%, respectively. However, DMSO was about 81% and 86.5%, much larger than DHF. From this tendency, it was predicted that the polarity of the solvent changed much more with the increase in $y_{w,0}$ in the case of DMF than DMSO.

Figure 1c,d, respectively, show the results of analysis of the raffinate recovered under constant conditions ($y_{w,0}$ = 0.03, $(S/F)_0$ = 10, T_0 = 28 °C) using DMF and DMSO. The gas chromatogram of raffinate is very different from that of the raw material (Figure 1b). It was found that the components, other than the paraffin components in the raw material, were extracted in large amounts by the extraction operation, and the peak of the paraffin component was remarkably increased, so that the concentration of the paraffin component in the raffinate increased. The concentrations of C_{12} and C_{16} paraffin components in the raffinate from $x_{i,P}$ of $y_{w,0}$ = 0 shown in Figure 2c increased approximately 1.12 and 1.56 times for DMF and about 0.98 and 1.15 times for DMSO, respectively, compared to their concentration in the raw material.

In the gas chromatogram of the raw material (distillate of WPPO) and the raffinate, each olefin component peak was positioned immediately before each paraffin component. It was expected that it would be difficult to separate the paraffin–olefin components by equilibrium extraction by comparing the gas chromatogram of the raw material (distillate of WPPO) in Figure 1b and the raffinate of Figure 1c.

Although this study only mentions the separation of paraffin components among raw materials by equilibrium extraction, we plan to conduct a detailed review of the separation of olefin components among raw materials by equilibrium extraction in the future.

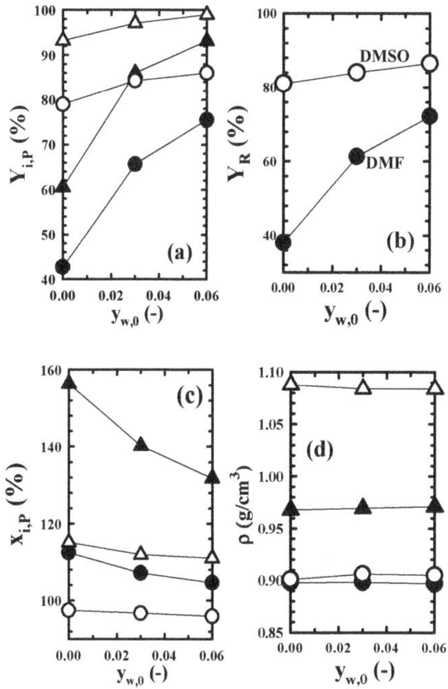

Figure 2. Comparison of solvent for (**a**) recovery rate of paraffin component i ($Y_{i,P}$) present in raffinate, (**b**) recovery rate of raffinate (Y_R), (**c**) the concentration increase rate of the paraffin component i ($x_{i,P}$) present in raffinate to that present in raw material (WPPO distillate), and (**d**) density (ρ) of raffinate and extract phase according to mass fraction of water in solvent of initial state ($y_{w,0}$). Experimental conditions: t = 72 h, $(S/F)_0$ = 3, T_0 = 28 °C. Keys of (**a**) and (**c**): • C_{12} (DMF), ▲ C_{16} (DMF), ○ C_{12} (DMSO), △ C_{16} (DMSO). Keys of (**d**): • raffinate phase (DMF), ○ raffinate phase (DMSO), ▲ extract phase (DMF), △ extract phase (DMSO).

In the extraction operation, the difference in density between two phases (raffinate phase, extract phase) is an important factor that determines the extraction processing speed. Figure 2d shows the density (ρ) of the two phases recovered after the extraction operation using DMF and DMSO. The effect of $y_{w,0}$ on the density of the raffinate and extraction phases was not recognized regardless of the solvent. The difference in the density of the two phases due to DMSO extraction was greater than that of DMF extraction. From this result, it was found that DMSO is more advantageous than DMF in terms of extraction processing speed.

When referring to the change in the color of the raffinate according to $y_{w,0}$ and the solvent, as shown in Figure 3, the color of the raffinate became lighter as the $y_{w,0}$ decreased regardless of solvent, and changed to very light yellow at $y_{w,0} = 0$ (raw material: dark brown). At the same $y_{w,0}$, the color of the raffinate recovered using DMF was lighter than that of DMSO. In particular, the raffinate recovered at $y_{w,0} = 0$ using DMF showed a pale yellow color, almost similar to the color of commercial light oil. Considering that the paraffin component is almost colorless, it can be confirmed that the concentration of paraffin components in the raffinate increases with decreasing $y_{w,0}$ in the case of DMF rather than DMSO.

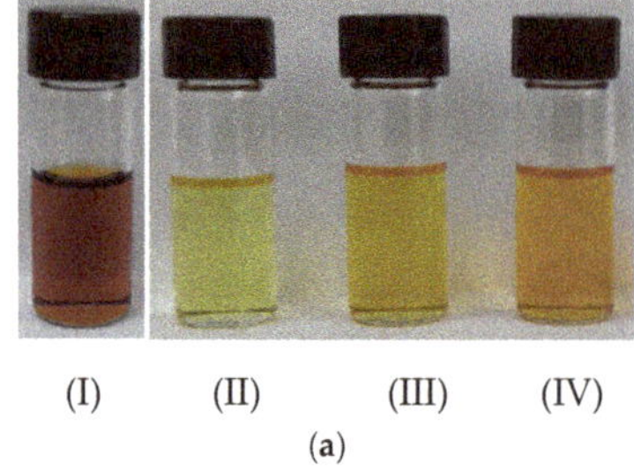

(I) (II) (III) (IV)

(**a**)

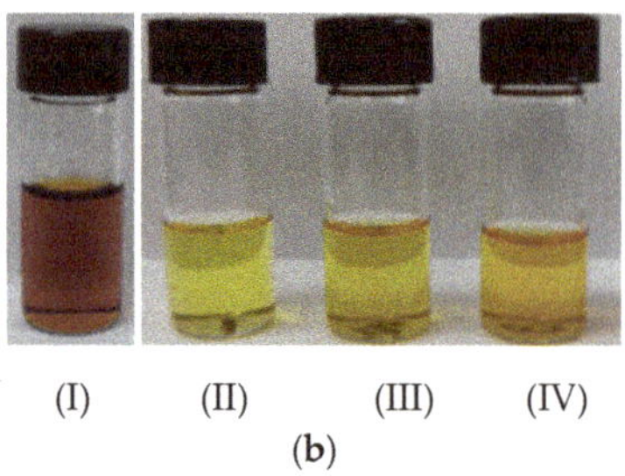

(I) (II) (III) (IV)

(**b**)

Figure 3. Comparison of solvent for color change in raffinate according to $y_{w,0}$. Solvent used: (**a**) DMF, (**b**) DMSO. Experimental condition: t = 72 h, $(S/F)_0 = 3$, $T_0 = 28$ °C. (I) raw material, (II) $y_{w,0} = 0$, (III) $y_{w,0} = 0.03$, and (IV) $y_{w,0} = 0.06$.

Furthermore, because the waxing phenomenon of commercial light oil can occur at a very low temperature, the raw material of this study and the raffinate recovered by DMF and DMSO extraction at $y_{w,0} = 0$ were left at −25 °C for 24 h. The occurrence of a waxing phenomenon was confirmed throughout this process. The waxing phenomenon was observed in the raw material and the raffinate recovered from DMSO, but not in the raffinate recovered from DMF.

Figure 4a shows that the extraction amount of paraffinic component *i*, which is extracted in the extraction phase, increased as $(S/F)_0$ increased regardless of the solvent used, while $Y_{i,P}$ decreased. In the case of $(S/F)_0 = 1$ for DMF, $Y_{i,P}$ of paraffin components C_{12} and C_{16} were about 84.9% and 100%, respectively, whereas those of DMSO were about 88.8% and 100%, respectively. Regardless of solvent, considering that the $Y_{i,P}$ of the C_{16} paraffin component is 100% at $(S/F)_0 = 1$, it was found that the paraffin components with a carbon number higher than 16 contained in the raw material remained at 100% in the raffinate recovered from $(S/F)_0 = 1$.

The Y_R values at $(S/F)_0 = 1$ for DMF and DMSO were about 81% and 88.6%, respectively (Figure 4b). Y_R decreased sharply as $(S/F)_0$ increased due to an increase in the extraction amount of all the components other than the paraffin components and the paraffin components contained in the raw material. At $(S/F)_0 = 10$, the Y_R values of DMF and DMSO were about 23.8% and 76.2%, respectively.

Figure 4c shows that $x_{i,P}$ increases with increasing $(S/F)_0$, regardless of the solvent used. This is considered to be a phenomenon caused by an increase in the selectivity of all the paraffin components compared to all other components such as aromatic components contained in the raw material by the increase in $(S/F)_0$. The concentration of C_{12} and C_{16} paraffin components in raffinate recovered at $(S/F)_0 = 10$ increased by about 1.37 and 2.46 times for DMF, 1.05 and 1.26 times for DMSO, respectively, compared to the concentration in the raw material. It is predicted that when DMF is used rather than DMSO, the selectivity of all the paraffin components to all the components other than the paraffin components contained in the raw material is large.

Figure 4d shows the density (ρ) of two phases recovered after the extraction operation using DMF and DMSO. Regardless of the solvent used, the effects of $(S/F)_0$ on the densities of the raffinate phase and the extract phase were not acceptable.

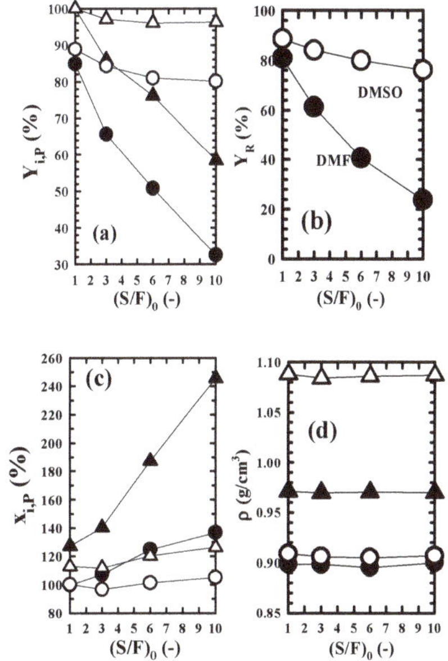

Figure 4. Comparison of solvent for (**a**) $Y_{i,P}$, (**b**) Y_R, (**c**) $x_{i,P}$, and (**d**) ρ according to mass ratio of solvent to raw material in initial state $(S/F)_0$. Experimental conditions: t = 72 h, $y_{w,0} = 0.03$, $T_0 = 28$ °C. Keys are shown in Figure 2.

Regardless of the solvent in Figure 5, it was found that the color of the raffinate becomes thinner as the $(S/F)_0$ increases, and becomes very pale yellow at $(S/F)_0 = 10$. DMF showed a color lighter than DMSO in the color of raffinate recovered at the same $(S/F)_0$ and, in particular, the raffinate recovered at $(S/F)_0 = 10$ using DMF showed a pale yellow color almost similar to the color of commercial light oil. As described above, when the raffinate obtained at $(S/F)_0 = 10$ was left at −25 °C for 24 h, a waxing phenomenon was observed in the raffinate recovered from DMSO, but not in that recovered from DMF.

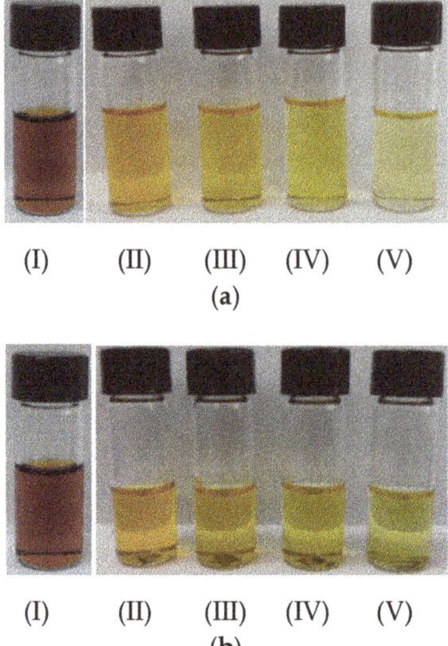

(I) (II) (III) (IV) (V)
(**a**)

(I) (II) (III) (IV) (V)
(**b**)

Figure 5. Comparison of solvent for color change in raffinate according to $(S/F)_0$. Solvent used: (**a**) DMF, (**b**) DMSO. Experimental conditions: $t = 72$ h, $y_{w,0} = 0.03$, $T_0 = 28$ °C. (I) raw material, (II) $(S/F)_0 = 1$, (III) $(S/F)_0 = 3$, (IV) $(S/F)_0 = 6$ and (V) $(S/F)_0 = 10$.

To confirm the reproducibility of all the experimental data obtained using two solvents in this study, the experiments were conducted under the same experimental conditions two or three times. The range of the reproducibility for the measured experimental value was within ± 5%.

Considering the effects according to $y_{w,0}$ and $(S/F)_0$ for the enrichment performance of paraffin components present in the raffinate, the color of the raffinate compared to commercial light oil and the waxing phenomena of the raffinate, DMF extraction was thought to be more advantageous from the point of view of the quality improvement of the distillate of WPPO than DMSO extraction. Of course, it is believed that the final evaluation of the products (raffinate) obtained from this study should be carried out in more detail through a quality evaluation according to the Petroleum Quality Inspection Method, as well as a review of the possibility of the separation of the olefin components contained in the distillate recovered from WPPO by DMF equilibrium extraction, which will be performed later.

Figure 6 shows the enrichment process of paraffin components from the distillate of WPPO, which is considered to be possible using the equilibrium extraction data of DMF. The proposed process was composed of one extraction tower (tower 1) to recover the paraffin components in the distillate of WPPO, one stripping tower (tower 2) to recover the extract in the extract phase, one washing tower (tower 3) to recover DMF in the raffinate and one distillation tower (tower 4) to recover the extraction solvent (DMF). This process is considered to be a valuable process for the enrichment of paraffin components contained in the distillate of WPPO.

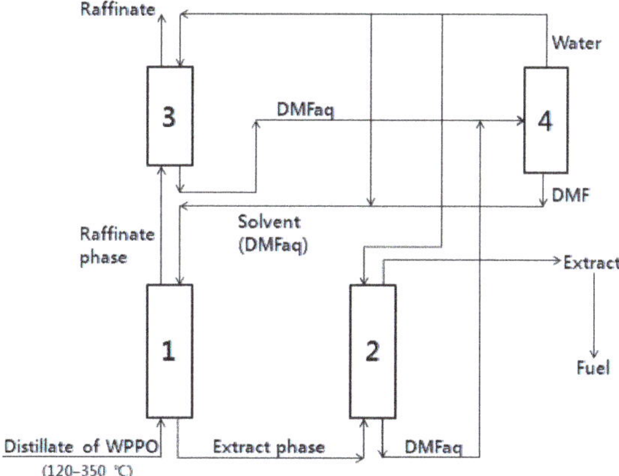

Figure 6. Enrichment process of paraffin components in distillate of WPPO. Tower no. 1: extraction tower, Tower no. 2: stripping tower, Tower no. 3: washing tower, Tower no. 4: distillation tower.

4. Conclusions

As part of the upgrading of the distillate (distilling temperature 120–350 °C.) of WPPO, this study examined the enrichment of paraffin components in the distillate by the equilibrium extraction of DMF and DMSO. Considering the effects according to $y_{w,0}$ and $(S/F)_0$ for the enrichment performance of paraffin components present in the raffinate, the color of the raffinate compared to commercial light oil and the waxing phenomena of the raffinate, DMF extraction was thought to be more advantageous from the point of view of the quality improvement of the distillate of WPPO than DMSO extraction. The process proposed using the results of DMF equilibrium extraction in this study, which comprised of one extraction tower, one stripping tower, one washing tower and one distillation tower, is a valuable process to enrich of the paraffin compounds contained in the distillate of WPPO.

Funding: This research received no external funding.

Conflicts of Interest: The author declare no conflict of interest.

References

1. Syamsiro:, M.; Saptoadi, H.; Norsujianto, T.; Noviasri, P.; Cheng, S.; Alimuddin, Z.; Yoshikawa, K. Fuel oil production from municipal plastic wastes in sequential pyrolysis and catalytic reforming reactors. *Energy Procedia* **2014**, *47*, 180–188. [CrossRef]
2. Kalargaris, I.; Tian, G.; Gu, S. The utilization of oils produced from plastic waste at different pyrolysis temperatures in a DI diesel engine. *Energy* **2017**, *131*, 179–185. [CrossRef]
3. Hartulistiyosoa, E.; Sigiroa, F.A.; Yuliantoa, M. Temperature distribution of the plastics pyrolysis process to produce fuel at 450 °C. *Procedia Environ. Sci.* **2015**, *28*, 234–241. [CrossRef]
4. Santaweesuk, C.; Janyalertadun, A. The production of fuel oil by conventional slow pyrolysis using plastic waste from a municipal landfill. *IJESD* **2017**, *8*, 168–173. [CrossRef]
5. Olufemi, A.S.; Olagboye, S.A. Thermal conversion of waste plastics into fuel oil. *Int. J. Petrochem. Sci. Eng.* **2017**, *2*, 252–257. [CrossRef]
6. Kalargaris, I.; Tian, G.; Gu, S. Experimental evaluation of a diesel engine fuelled by pyrolysis oils produced from low-density polyethylene and ethylene–vinyl acetate plastics. *Fuel Process. Technol.* **2017**, *161*, 125–131. [CrossRef]
7. Frigo, S.; Seggiani, M.; Puccini, M.; Vitolo, S. Liquid fuel production from waste tyre pyrolysis and its utilization in a diesel engine. *Fuel* **2014**, *116*, 399–408. [CrossRef]
8. Hamidi, N.; Tebyanian, F.; Massoudi, R.; Whitesides, L. Pyrolysis of household plastic wastes. *Br. J. Appl. Sci. Technol.* **2013**, *3*, 417–439. [CrossRef]

9. Ratnasari, D.K.; Nahil, M.A.; Williams, P.T. Catalytic pyrolysis of waste plastics using staged catalysis for production of gasoline range hydrocarbon oils. *J. Anal. Appl. Pyrolysis* **2017**, *124*, 631–637. [CrossRef]
10. Sharratt, P.N.; Lin, Y.H.; Garforth, A.A.; Dwyer, J. Investigation of the catalytic pyrolysis of high-density polyethylene over a HZSM-5 catalyst in a laboratory fluidized-bed reactor. *Ind. Eng. Chem. Res.* **1997**, *36*, 5118–5124. [CrossRef]
11. Bagri, R.; Williams, P.T. Catalytic pyrolysis of polyethylene. *J. Anal. Appl. Pyrolysis* **2002**, *63*, 29–41. [CrossRef]
12. Lee, K.H.; Shin, D.H. A comparative study of liquid product on non-catalytic and catalytic degradation of waste plastics using spent FCC catalyst. *Korean J. Chem. Eng.* **2006**, *23*, 209–215. [CrossRef]
13. Koca, A.; Bilgesub, A.Y. Catalytic and thermal oxidative pyrolysis of LDPE in a continuous reactor system. *J. Anal. Appl. Pyrolysis* **2007**, *78*, 7–13. [CrossRef]
14. Wiriyaumpaiwong, S.; Jamradloedluk, J. Fast pyrolysis of plastic wastes. *Energy Procedia* **2017**, *138*, 111–115. [CrossRef]
15. Murugan, S.; Ramaswamy, M.C.; Nagarajan, G. The use of tyre pyrolysis oil in diesel engines. *Waste Manag.* **2008**, *28*, 2743–2749. [CrossRef]
16. Kumar, R.; Mishra, M.K.; Singh, S.K.; Kumar, A. Experimental evaluation of waste plastic oil and its blends on a single cylinder diesel engine. *J. Mech. Sci. Technol.* **2016**, *30*, 4781–4789. [CrossRef]
17. Kim, S.J.; Kim, S.C. Separation and recovery of dimethylnaphthalene isomers from light cycle oil by distillation-extraction combination. *Sep. Sci. Technol.* **2003**, *38*, 4095–4116. [CrossRef]
18. Kim, S.J. Purification of 2,6-dimethylnaphthalene containing in light cycle oil by distillation-solvent extraction-solute crystallization combination. *J. Ind. Eng. Chem.* **2019**, *79*, 146–153. [CrossRef]
19. Jiao, T.T.; Zhuang, X.L.; He, H.Y.; Li, C.S.; Chen, H.N.; Zhang, S.J. Separation of phenolic compounds from coal tar via liquid liquid extraction using amide compounds. *Ind. Eng. Chem. Res.* **2015**, *54*, 2573–2579. [CrossRef]
20. Jiao, T.T.; Li, C.S.; Zhuang, X.L.; Cao, S.S.; Chen, H.N.; Zhang, S.J. The new liquid-liquid extraction method for separation of phenolic compounds from coal tar. *Chem. Eng. J.* **2015**, *266*, 148–155. [CrossRef]
21. Kang, H.C.; Kim, S.J. Comparison of methanol with formamide on separation of nitrogen heterocyclic compounds from model coal tar fraction by batch cocurrent multistage equilibrium extraction. *Polycycl. Aromat. Compd.* **2016**, *36*, 745–757. [CrossRef]
22. Kim, S.J.; Chun, Y.J. Separation of nitrogen heterocyclic compounds from model coal tar fraction by solvent extraction. *Sep. Sci. Technol.* **2005**, *40*, 2095–2109.
23. Kim, S.J. Separation and purification of indole in model coal tar fraction of 9 compounds system. *Polycycl. Aromat. Compd.* **2019**, *39*, 60–72. [CrossRef]
24. Kim, S.J.; Kim, S.C. Separation of valuable bicyclic aromatic components from light cycle oil by an emulsion liquid membrane. *Sep. Sci. Technol.* **2004**, *39*, 1093–1109. [CrossRef]
25. Egashira, R.; Habaki, H.; Kawasaki, J. Decrease in aromatics content in motor gasoline by O/W/O emulsion liquid membrane process. *J. Jpn. Pet. Inst.* **1997**, *40*, 107–114. [CrossRef]
26. Kim, S.J.; Kang, H.C.; Kim, Y.S.; Jeong, H.J. Liquid membrane permeation of nitrogen heterocyclic compounds contained in model coal tar fraction. *Bull. Korean Chem. Soc.* **2010**, *31*, 1143–1148. [CrossRef]
27. Sharma, A.; Goswami, A.N.; Rawat, B.S.; Krishina, R. Effect of surfactant type on the selectivity for separation of 1–methyl naphthalene and dodecane using liquid membranes. *J. Membr. Sci.* **1987**, *42*, 19–30. [CrossRef]
28. Kang, H.C.; Shin, S.S.; Kim, D.H.; Kim, S.J. Recovery of paraffin components from pyrolysis oil fraction of waste plastic by batch cocurrent 4 stages equilibrium extraction. *Appl. Chem. Eng.* **2018**, *29*, 630–634.

Article

European Union Legislation Overview about Used Vegetable Oils Recycling: The Spanish and Italian Case Studies

Jesus Ibanez [1], Sonia Martel Martín [1], Salvatore Baldino [2], Cristina Prandi [2] and Alberto Mannu [2,*]

1 International Research Centre in Critical Raw Materials (ICCRAM), University of Burgos, Plaza Misael Bañuelos s/n, 09001 Burgos, Spain; jesusibanez@ubu.es (J.I.); smartel@ubu.es (S.M.M.)

2 Department of Chemistry, University of Turin, Via Pietro Giuria, 7, I-10125 Torino, Italy; salvatore.baldino@unito.it (S.B.); cristina.prandi@unito.it (C.P.)

* Correspondence: alberto.mannu@unito.it

Received: 20 June 2020; Accepted: 6 July 2020; Published: 8 July 2020

Abstract: The employment of used vegetable oils (UVOs) as raw materials in key sectors as energy production or bio-lubricant synthesis represents one of the most relevant priorities in the European Union (EU) normative context. In many countries, the development of new production processes based on the circular economy model, as well as the definition of future energy and production targets, involve the utilization of wastes as raw material. In this context, the main currently applied EU regulations are presented and discussed. As in the EU, the general legislative process consists of the definition in each State Member of specific legislation, which transposes the EU indications. Two relevant countries are herein considered: Italy and Spain. Through the analysis of the conditions required in both countries for UVOs' collection, disposal, storage, and recycling, a wide panorama of the current situation is provided.

Keywords: used vegetable oil; recycling; biodiesel; bio-lubricant; legislation

---

1. Introduction

The development of new sustainable processes based on the circular economy model is one of the main targets of the European Union (EU), according to the Circular Economy Action Plan published in March 2020 [1]. To reach such a goal, a combination of several factors is needed. Opportune raw materials arising from recycling processes must be chosen and transformed by low-impacting steps where near-zero wastes are produced. Within the most promising candidates for this approach, used vegetable oils (UVOs) have shown excellent characteristics. In fact, UVOs are wastes produced worldwide continuously in large amounts [2], and they are easily recyclable. A very recent overview covering the current available processes and technologies for the recycling of waste cooking oils (WCOs), which represent the main source of UVOs, has been reported by Mannu et al. [3]. In general, UVOs can be employed as raw materials for many industrial productions [4,5]. The possible applications of UVOs include the production of bio-lubricants [6–9], biofuels [10–13], energy [14,15], animal feeds [16,17], ecological solvents [18], composites materials [19,20], and non-aqueous gas sorbents devices [21,22], only to cite the most relevant.

The amount of UVOs produced has a great impact. It can be estimated that around 200 million tons per year, either from food or industrial sectors, are produced [23].

The increasing use of UVOs for the production of biodiesel, which adsorbs nearly 90% of the UVOs raw material, has grown a 6 billion dollar market, and it is expected to reach 8.88 billion dollars in 2026.

In the European Union (EU), according to the European Biomass Industry Association (EUBIA), 4 million tons of UVOs are produced every year, and only one-seventh of this amount is collected [24].

In this context, the technical and economic advantages associated with the employment of UVOs as raw materials are accompanied by a strong legislative endorsement. As a matter of fact, many countries have developed specific legislation to favour the UVOs' transformation in added-value goods.

Herein, the general European Union legislative framework regarding the aforementioned topic will be described. An overview of the currently most relevant EU legislation dealing with UVOs will be provided, followed by a case-study detailed analysis of the matter in two key Member States: Italy and Spain.

2. Discussion

2.1. European Framework for WCOs' Collection and Recycling

As expected, UVOs' disposal and recycling have been the object of discussion and legislative decisions in the European Union (EU) only in very general terms. Actually, the scope of the EU bodies is precisely to promulgate guidelines and general legislation, which can be taken as reference from each Member State to properly develop local laws, and the case of UVOs is not an exception.

Used vegetable oils fall into the formal category of "Edible Oil and Fats", which is part of the waste family Municipal Wastes (household waste and similar commercial, industrial, and institutional wastes, including separately collected fractions). The specific description of this category can be found in the Commission Decision 2014/955/EU [25] of 18 December 2014, which amends Decision 2000/532/EC (EC stands for European Commission) on the list of waste, pursuant to Directive 2008/98/EC. The main purpose of the document is to assess the risk for the public health associated with waste by defining specific cut-off values for selected recognized dangerous chemicals. The employed criteria for hazardousness determination are reported in Annex III to Directive 2008/98/EC [26]. Concerning the category edible oils and fats, they are classified in Decision 2000/532/EC at the number 200125 [27]. In this context, the EU Directive 2018/851 goes one step forward, defining some crucial aspects related to the life cycle of wastes, including UVOs [28].

In particular, within the main aims reported in the introductory section of the Directive, the following indications stand out.

(I) A special effort must be pointed toward the enhancement in actions aimed to improve the sustainable transformation of waste-based raw materials, possibly under circular economy models.
(II) Specific actions must be considered to improve the recycling and re-use of wastes.
(III) Waste oils must be collected in an exclusive way in order to facilitate their treatment and recycling.
(IV) Concerning the treatment of waste oils, their transformation into added-value products must be preferred with respect to other transformations (e.g., destruction).

In particular, to push Member States toward actions pointed to pursuit the abovementioned points, Article 21 of the Directive 2008/98/EC was amended by adding the following paragraph: "*4. By 31 December 2022, the Commission shall examine data on waste oils provided by Member States in accordance with Article 37(4) with a view to considering the feasibility of adopting measures for the treatment of waste oils, including quantitative targets on the regeneration of waste oils and any further measures to promote the regeneration of waste oils. To that end, the Commission shall submit a report to the European Parliament and to the Council, accompanied, if appropriate, by a legislative proposal.*"

The directive 2004/35/CE [29], which "*establish a common framework for the prevention and remedying of environmental damage at a reasonable cost to society*" (article 3), furnishes specific indications to the Member States on the responsibility of the operators towards the environmental damages produced by the incorrect disposal of dangerous wastes, including oil pollution damage (lubricant and edible). The specific environmental risk related to oil pollution uncontrolled disposal is referred to protected species and natural habitats, water and land damage. The directive addresses different life cycle stages of vegetable oil, from their disposal, once exhausted, to their use, once recycled.

For example, in the case of bio-lubricants for chains, they are dispersed into the environment during use, and, thus, their potential impact on the former aspects has to be taken into account.

Summarizing this general context, waste edible oils, which include vegetable oil and fats at the end of their life cycle, must be valorized through their use as raw materials for the production of added-value products, possibly by adopting a circular economy model, which by definition reduces the environmental impact. To reach such ambitious targets, some scientific and technological efforts need to be made. In this regard, the currently known technologies able to process waste vegetable oils as raw materials matching with the circular economy model are few and sometimes limited by the available technologies and materials [3]. This reflects into the corresponding legislation, and there are few processes involving waste vegetable oils that have been the object of specific attention by the legislator. The Commission Regulation 142/2011 of 25 February 2011 [30], which implements the Regulation (EC) No 1069/2009 of the European Parliament and of the Council, deals with the processing of several raw materials, including catering wastes. Specific applications are considered, such as incineration, animal feed, biodiesel, biogas, and composting. In addition, the regulation describes the requisites needed for market uptake and internationalization.

Nearly 90% of UVOs collected are processed for energy purposes and, in particular, as raw materials for biodiesel production [31]. Energy is a driving sector in the EU, and the legislation related is relevant. The Directive (EU) 2018/2001 of the European Parliament and of the European Council of 11 December 2018, known as the Renewable Energy Directive (RED II), discusses the valorization for biodiesel production. RED II extends the objectives and the application period of the previous RED I, dated 2009. In particular, it increases, by 2030, the targets for the use of renewable energies to 32% while setting to 14% the quote of fuels consumed by the transport sector.

RED II includes the adoption of new actions to promote a second generation of biofuels to be exploited in the transport sector, pushing towards a reduction in the CO_2 emissions and in favour of the adoption of circular economy models. As a consequence of that, in the near future, a consistent and increasing percentage of fuel will arise by biomass, including UVOs.

The main instrument for assessing the progress towards the aforementioned targets established in RED I and reinforced in RED II is the double-counting rule, which allows counting twice in the calculation of the shares of renewable energies, the waste-based biofuels. This system promotes the consumption of second-generation waste-based biofuels produced, e.g., from feedstocks as UVOs.

Regarding the specific routes to achieve the abovementioned targets, they are taken into account in the regulation "Action for Climate" 2018/1999. Each EU Country should develop a dedicated national renewable energy plan according to this general regulation [32,33].

In addition, the European Union launched in 2015 the First Circular Economy Action Plan, a strategy to start the way to improve the valorization and utilization of waste in the community [34], followed by the new Action Plan *"A new Circular Economy Action Plan For a Cleaner and more Competitive Europe"*, framed in the European "Green Deal". It arises from the fact that the EU considers the development of actions and synergies to enhance the adoption of circular economy models in several sectors as a priority activity. The general target of such actions is to maximize the natural resources use, thus reducing waste and consequent pollution from mankind's activities. In the current version of the Green Deal, it is expected that the EU will reach carbon neutrality by 2050. To reach such a target, the demand for specific raw materials, including waste vegetable oils as well as concrete, steel, plastic, glass, aluminum, copper, and iron, and in general technology-specific materials, is expected to increase in the following years. In particular, the development of wind and solar technology will generate the need to import extra-EU massive amounts of many raw materials, increasing their market, which will need specific further regulations [35].

The first circular economy plan was presented in 2015, and it consisted of 54 actions to be pursued by the EU Commission during the period 2015–2020. These actions were directed to the optimization of the main steps of the life cycle of goods (design, production, consumption, management of wastes, and recycling). Furthermore, some sectorial aspects were treated, such as those related to plastic packaging, food wastes, critical raw materials (CRMs), wastes from constructions, and biomass, considered pivotal for sustainable development.

The new 2020 Plan aims to enhance the circularity, presenting specific procedures in key sectors, such as electronics, batteries, automotive, packaging, plastics, textiles, constructions, and food. The minimization of the employment of natural raw materials and the production of wastes is also discussed. An improvement in the secondary market of raw materials and a consistent reduction in the pollution by 2050 are presented as the main targets. This action plan is further developed for the food sector in the "From Farm to Fork Strategy", *"A Farm to Fork Strategy for a fair, healthy and environmentally-friendly food system"* published in May 2020 [36], where the reduction in food loss and waste are key points to achieve the goals of the strategy. The policy in this regard will focus on the recovery of nutrients and secondary raw materials, the production of feed, food safety, biodiversity, bioeconomy, waste management, and renewable energy in the upcoming years.

Another relevant normative expression concerning UVOs is Regulation (EC) No. 1069/2009 of the European Parliament and of the Council, also known as the *"Animal by-products Regulation"*, which deals with the management of animal by-products for non-food applications. The main target of this regulation is decreasing the risks to public and animal health by enhancing the quality of the food chain. During the frying process, the chemical composition of vegetable oil is subjected to many transformation processes, many producing dangerous molecules. As a matter of fact, UVOs' recycling and its re-utilization as animal feed, has the effect of re-introducing into the food chain potentially harmful chemicals. For this reason, a careful and specific characterization, regulated by the specific legislation, must be pursued before the processing of UVOs as raw material for animal food additives [37].

2.2. *Italy*

Currently, the yearly production of UVOs in Italy can be estimated at 260 K tons, partitioned between the food industry (36%) and household kitchens (64%). Authorized collectors were able to recover 72 K tons in 2017, corresponding to 27% of the total estimated [38].

From a normative point of view, the disposal and management of wastes in Italy are regulated by Decree (D. lgs.) 152/06 [39]. The EU directives, and in particular the ones contained in the *"Waste Framework Directive* 98/2008/CE" were introduced in Italy in 2010, with the D. lgs. 205/2010 [40], which adds some modifications to D. lgs 152/06.

Among the additional indications furnished through this legislation, of primary importance are the management of collection and treatment (by recycling or disposal) of used vegetable oils and animal fats through a dedicated consortium for the collection and treatment of the used vegetable oils (CONOE). Actually, the CONOE was already named in Article 47 of the Decree D. lgs. 22/97, modified by Article 233 of the D. lgs. 152/06.

The main tasks and targets of the CONOE are the followings:

(I) To manage and develop, in accordance with the European and National legislations, the systems of collection, transport, storage, treatment, and recycling of used oils and fats.
(II) When recycling processes cannot be performed, to guarantee the proper disposal procedures. The main aspect considered in the evaluation of the convenience of disposal versus recycling is the sustainability of the entire process, especially in terms of the production of undesired pollutants.
(III) To manage all the aspects related to the collection, transport, storage, recycling, or disposal of used vegetable oils and fats, to minimize the pollution eventually generated by this waste, the consortium represents a common point of contact for all the companies involved in such aspects. With the target to promote the dissemination of relevant information and to enhance cooperativity among the partners, the CONOE develops activities as market analyses and surveys, publish economic and technical reports, and organizes transversal activities within the partners.

The mechanism of the management of UVOs developed by the CONOE can be summarized as follows. I) Waste producers affiliate to the consortium directly or through the corresponding category association. II) UVOs producers deliver the waste to a collector associated with the consortium, who will deliver the oil to a processing company that must be associated with the CONOE, as well.

From a practical point of view, all the actors involved in the life cycle of UVOs are associated with the national consortium and, thus, are subjected to its monitoring activity. In 2018, the CONOE was constituted by more than 300 K producers, more than 450 companies dealing with collection and storage, more than 60 recycling industries [38].

Despite the many possible applications of UVOs, the main destination of this raw material is represented by the biodiesel industry. Currently, in Italy, 90% of the collected UVOs is processed in a biodiesel plant. The interest toward enhancing biodiesel production from renewable resources (such as UVOs) is based on two different aspects: from one side, the technology needed for such production is already developed and does not require any variation when UVOs are introduced as raw materials. From the other side, EU legislation, such as the Directive RED (Renewable Energy Directive) 2009/28/CE and FQD (Fuel Quality Directive) 2009/30/CE [40], introduced sustainability as the main parameter to consider the fuels produced as effective for the count of the renewable energy targets (vide infra).

Nevertheless, the potential of UVOs for non-biodiesel applications is slowly emerging as confirmed by the recent development of large scale biorefineries based on UVOs and dedicated, in part, to the production of bio-lubricants [41,42].

2.3. Spain

As for Italy, the Spanish legislation related to the recycling of UVOs is derived from the mandatory adoption of the EU energy and waste policies, which have been transcribed into national laws by each Member State. Moreover, UVOs' collection and management are directly related to the objectives established by the different Spanish local governments for the use of biofuels and renewable energy (Figure 1).

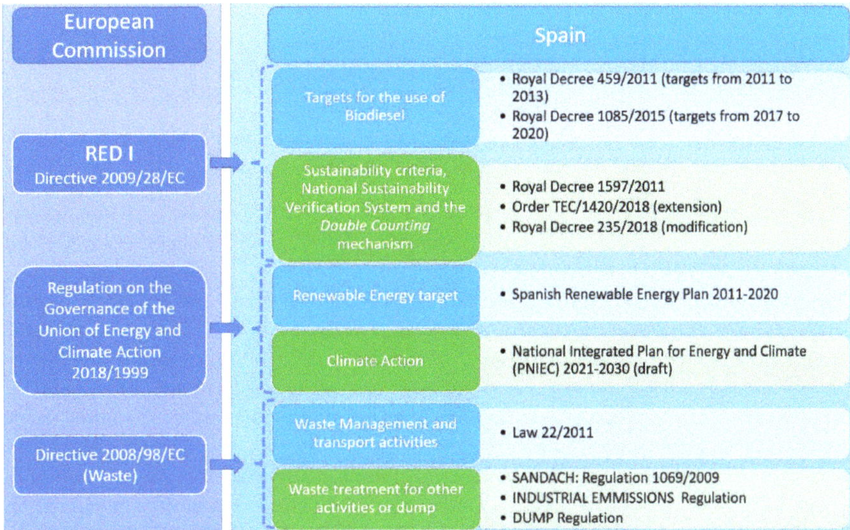

Figure 1. European Union (EU) regulation transposed into Spanish legislation. EC stands for European Commission, and RED for Renewable Energy Directive.

The Royal Decree 459/2011 [43] follows the instructions of the RED I Directive and set mandatory targets for the use of biofuels for 2011, 2012, and 2013, expressed as minimum energy content, in relation to the energy content of petrol, diesel, and total petrol and diesel sold or consumed.

The original targets for biofuels in total, biofuels in diesel, and biofuels in petrol were then extended by Royal Decree 1085/2015 to establish specific targets for the years 2017 (5%), 2018 (6%), 2019 (7%,) and 2020 (8.5%) [44].

The current use of biofuels is difficult to calculate due to the controversy over the methodology in accounting for the data reported by the biodiesel producers, but it is estimated that in 2019, the use was 6%, and the objective for the mentioned year (7%) could be reached in the final annotation, which will be published in the following months [45].

Royal Decree 1597/2011 [46] regulates the sustainability criteria for biofuels and bioliquids. The National Sustainability Verification System and the Double Counting mechanism regulates the calculation of biofuels' production from renewable sources. This Royal Decree encourages the production of second-generation biofuels obtained from wastes, residues, non-food cellulosic material, and lignocellulosic material by imposing their energy contribution as the double of the one relative to other biofuels. This criterium was already described in the RED I normative (*Double Counting*). Royal Decree 1597/2011 was then extended in the Order TEC/1420/2018 [47] and modified by Royal Decree 235/2018 [48].

In addition, the Spanish Renewable Energy Plan 2011–2020 sets targets in line with Directive 2009/28/EC (RED I). The new National Integrated Plan for Energy and Climate (PNIEC) 2021–2030 [49] is in draft status and was sent to the European Union in January 2020, complying with the indications established in the regulation on the Governance of the Union of Energy and Climate Action 2018/1999. It is intended to influence the current business models, society, and economy in general, by pursuing new investments and by increasing job demand. The primary target of the Energy and Climate Action is to reduce greenhouse gas emissions, pushing Spain towards climate neutrality by 2050, accomplishing the requirements of the Paris Agreement. Other priorities include the promotion of energy rehabilitation in buildings, the promotion of energy self-consumption by citizens, and encouraging biomasses utilization for renewable energy production.

In this context, the main specific objectives are:

- 23% reduction in greenhouse gas (GHG) emissions with respect to 1990,
- 42% of renewables in energy end-use. This figure is double the 20% of the year 2020,
- 39.5% improvement in energy efficiency over the next decade,
- 74% presence of renewable energies in the electricity sector, in coherence with a trajectory towards a 100% renewable electricity sector in 2050.

Other important aspects are considered in the law 22/2011 [50] on waste, which transposes the Directive 2008/98/EC. It regulates waste management and transport activities, as well as the authorizations required for waste treatment.

The accordance between the collection and transport of UVOs and law 22/2011 is established in Regulation 1069/2009, named in Spain as SANDACH (Subproductos Animales no Destinados al Consumo Humano), which states in article 21: "Operators shall collect, transport and dispose of category 3 catering waste in accordance with the national measures provided for in Article 13 of Directive 2008/98/EC." Collection always implies storage at a physical location. Therefore, storage in the area of UVOs collection must comply with the mentioned law [51].

Furthermore, whether the collectors use UVOs to produce biofuel, they have to comply with the provisions of the RED Directive, since it includes the specific definition for the manufacturing of biofuels. Traceability can be verified by the ISCC (International Sustainability and Carbon Certification), which guarantees that the collector carries out a tracking procedure for the raw waste materials (UVOs), and requires a self-declaration which assures the origin of the UVOs. Moreover, to comply with the RED Directive, the collectors should also confirm the authenticity of the waste by declaring that they did not produced it, e.g., by adding fresh oil to a used one. Companies involved in the whole chain can be audited to ensure compliance with the requirements of the regulations [52].

Therefore, a company dedicated to the collection and management of UVOs in Spain must comply with the above regulations, ensuring traceability so that the origin of the UVOs can be verified at any time and carry out collection, transport, storage, and treatment in accordance with the national waste law. Companies producing biodiesel must do so according to the methodology established by the SANDACH Regulation (Figure 2).

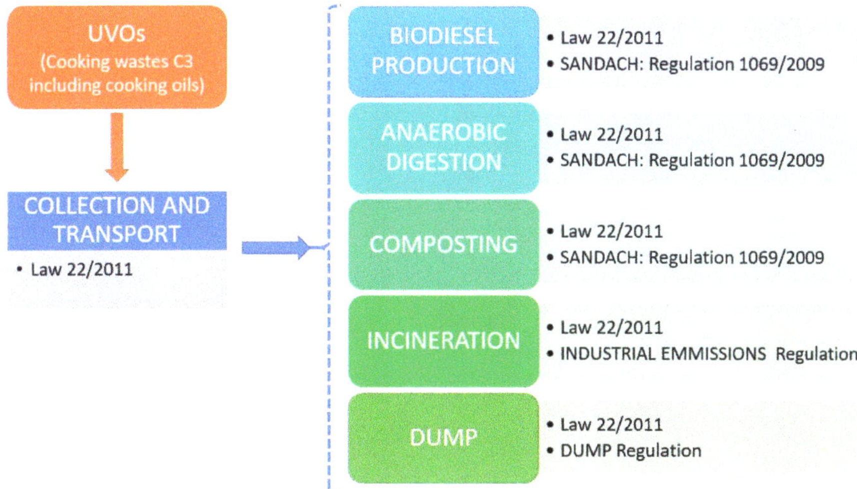

Figure 2. Overview of Spanish regulation involving used cooking oils' collection and treatment. UVOs stands for Used Vegetable Oils, SANDACH for Subproductos Animales no Destinados al Consumo Humano, and DUMP Regulation is referred to the regulation related to the dumping of wastes in the environment.

The Spanish equivalent of the Italian CONOE is the *National Association of Waste and Edible Oil and Fat By-Product Managers* (GEREGRAS), which was founded in 2007. It is constituted by several companies dedicated to the management of UVOs. The association provides advice on many aspects with the target of promoting the development of companies in the sector, and it is sometimes involved in negotiations with public institutions as advisor for the management of UVOs in the municipalities. GEREGRAS associates collect 70% of the UVOs generated in the Hotellerie-Restaurant-Catering (HORECA) sector and 80% of UVOs arising from domestic waste. The volume of UVOs collected in Spain is 36% of the total produced, a percentage quite far from the 60% target for 2030 indicated by GEREGRAS.

The current delay with respect to the expected level of UVOs' collection and recycling is mainly due to the low collection rate from the domestic sector (4%). It can be observed that the domestic sector experiences a lack of specific regulation that induces the citizens to contribute to UVOs' recycling actively. It is crucial to improve the household collection systems, making recycling easier for the citizens. In addition, governments play a very important role in terms of awareness of the population [53]. This awareness leads public and private institutions working on different mechanisms and awareness campaigns to ensure and facilitate the collection of UVOs from private houses, and, thus, be able to meet the objective of 60% in 2030 (Table 1) [54]. The competences in the management of waste relate to the local authorities according to Law 22/2011, and with it, awareness. Therefore, improving the amount of managed UVOs means coordination between local and regional governments and joint improvement, as there is a very high potential amount.

Table 1. Current data on used vegetable oils' (UVOs) collections in Spain and future outlooks [54].

	HORECA and Industrial	Household	Total
UVOs to be managed	160,000	178,900	338,900
UVOs currently managed	115,200	7300	122,500
Percentage managed	72%	4%	36%

The low amounts of UVOs collected reflect the not optimized household collection system, which makes the sector uncompetitive, producing an internal need for vegetable oil raw material, which is satisfied by importing waste oils from other countries to achieve the renewable energy targets.

An additional issue related to the fast-growing UVOs market between countries due to the high targets set by the EU, consists in a lack of traceability in the UVOs supply chain, making the origin of UVO sometimes not traceable. According to Farm Europe [55], the fraudulent UVO is that which does not come from its use in kitchens, but from virgin palm oil. This does not comply with the requirements of the RED Directive for the composition of biodiesel.

The renewable energy objectives established in RED I and extended in the new RED II, have caused a growth in the demand for UVO, being 2200 million liters in 2017, and estimating that it will exceed 3000 million liters in 2020. This demand cannot be met with the UVO currently collected, so 62% is imported, of which 75% comes from Asia. The difficulty of carrying out a correct traceability of the origin of the imported UVO raises doubts about their composition, as virgin palm oil can easily be added and sold as a UVO.

Farm Europe also states that monitoring the whole supply chain of UVO is a great challenge. Voluntary traceability systems in the EU have proven not to be enough to ensure the origin. They point out a curious fact from Taiwan, a country that, in 2017, treated 12,500 tonnes of its UVO, but exported 11 million tonnes to countries such as the United Kingdom (UK) and Ireland. The Double Counting Rule of the RED Directive, mentioned above, was created to encourage the collection system of UVO in Europe and, thus, generate growth in these businesses. However, the lack of control mechanisms has led to a drastic increase in imports and the benefits of biofuel production, with UVO, being taken away by third countries.

The International Sustainability and Carbon Certification (ISCC), the certification system for collecting companies mentioned above, has included points of improvement in traceability that will be applicable in 2020, and it is intended that they will improve some of the points mentioned.

Ensuring a robust traceability system is a key point to avoiding possible deviations in the value chain, from the production of cooking oil, its use, and its correct management once used. Furthermore, this also avoids the promotion of the underground economy. Currently, there are thefts of containers that have collected UVO by individuals who sell the stolen UVO to companies that do not give importance to the origin, causing an economic loss for management companies that invest in improving collection mechanisms, in addition to the loss of taxes by selling without invoicing.

It is absolutely necessary to improve the monitoring of the whole value chain of UVOs so that the biodiesel produced is truly sustainable and a second-generation biofuel, and to ensure that the regulations do not lose credibility. EWABA (European Waste-to-Advanced Biofuels Association), has established a new method for the physical differentiation test between used and virgin oils, which seems satisfactory and which is intended to be implemented in the short term [56,57].

3. Outlooks

As reported in the previous chapters, the focus at the normative level has been centered on the application of UVOs for energy purposes. This is mainly due to the already existing technology for biodiesel production, which indeed is easily adaptable to UVOs' processing. Nevertheless, emerging scientific, technological, and economic trends are moving the attention toward the need for new regulations, specifically in the fields of bio-lubricant production [58,59] and in the quality assessment of the UVO-based raw materials [60]. In particular, regarding the quality assessment of UVOs as raw materials, EU Member States are considered directly responsible for quality monitoring. Thus, each country should adopt a "voluntary scheme" to certificate the produced biofuel. Currently, this specific situation is evolving, and specific local legislations are the object of internal discussions.

4. Conclusions

The rise in used vegetable oil (UVOs) as a valuable feedstock for industry is being endorsed by the European normative framework. In particular, the versatility of this waste raw material, which can be easily processed through sustainable processes, even by adopting a circular economy model, has pushed many Member States to develop local normative and promote the recycling of UVOs. Nevertheless, the process of transposition of the EU normative into local legislation is not complete, and several aspects remain unresolved. Italy and Spain, two of the most representative countries of the EU, have adopted the EU indications, especially in terms of collection, disposal, and recycling for energy purposes, mostly for biodiesel production. Minor markets based on UVOs as bio-lubricants or animal feed production are treated by the EU and national legislations only marginally. On the other side, the increasing economy related to biodiesel production from vegetable oils has generated a huge demand for raw materials, which are sometimes required from other countries, even external to the EU. This recent tendency has not been the object of proper legislative action yet, and it represents a problem in terms of quality control of the employed raw materials.

Author Contributions: Conceptualization, A.M. and J.I.; methodology, A.M.; investigation, J.I. and A.M.; resources, A.M. and S.M.M.; data curation, S.M.M.; writing—original draft preparation, A.M. and J.I.; writing—review and editing, S.M.M., C.P., and S.B. All authors have read and agreed to the published version of the manuscript.

Funding: This research received no external funding.

Acknowledgments: The authors thank all of the people involved with the MSCA RISE 2019 project WORLD (873005).

Conflicts of Interest: The authors declare no conflict of interest.

References

1. European Commission. COM (2020) 98 Final, 11.3.2020 "A new Circular Economy Action Plan for a Cleaner and More Competitive Europe". Available online: https://ec.europa.eu/environment/circular-economy/ (accessed on 7 July 2020).
2. Available online: http://www.eubia.org/cms/wiki-biomass/biomass-resources/challenges-related-to-biomass/used-cooking-oil-recycling/ (accessed on 13 June 2020).
3. Mannu, A.; Garroni, S.; Ibanez Porras, J.; Mele, A. Available Technologies and Materials for Waste Cooking Oil Recycling. *Processes* **2020**, *8*, 366. [CrossRef]
4. Borrello, M.; Caracciolo, F.; Lombardi, A.; Pascucci, S.; Cembalo, L. Consumers' Perspective on Circular Economy Strategy for Reducing Food Waste. *Sustainability* **2017**, *9*, 141. [CrossRef]
5. Mannu, A.; Ferro, M.; Di Pietro, M.E.; Mele, A. Innovative applications of waste cooking oil as raw material. *Sci. Prog.* **2019**, *102*, 153–164. [CrossRef] [PubMed]
6. Mannu, A.; Ferro, M.; Colombo Dugoni, G.; Panzeri, W.; Petretto, G.L.; Urgeghe, P.; Mele, A. Improving the recycling technology of waste cooking oils: Chemical fingerprint as tool for non-biodiesel application. *Waste Manag.* **2019**, *96*, 1–8. [CrossRef]
7. Shashidhara, Y.M.; Jayaram, S.R. Vegetable oils as a potential cutting fluid-An evolution. *Tribol. Intern.* **2010**, *43*, 1073–1081. [CrossRef]
8. Zhang, Y.; Dube, M.A.; McLean, D.D.; Kates, M. Biodiesel production from waste cooking oil: Process design and technological assessment. *Bioresour. Technol.* **2003**, *89*, 1–16. [CrossRef]
9. Hamze, H.; Akia, M.; Yazdani, F. Optimization of biodiesel production from the waste cooking oil using response surfacemethodology. *Process. Saf. Environ. Prot.* **2015**, *94*, 1–10. [CrossRef]
10. Namoco, C.S., Jr.; Comaling, V.C.; Buna, C.C., Jr. Utilization of used cooking oil as an alternative cooking fuel resource. *ARPN J. Eng. Appl. Sci.* **2017**, *12*, 435–442.

11. Capuano, D.; Costa, M.; Di Fraia, S.; Massarotti, N.; Vanoli, L. Direct use of waste vegetable oil in internal combustion engines. *Renew. Sustain. Energy Rev.* **2017**, *69*, 759–770. [CrossRef]
12. No, S.Y. Inedible vegetable oils and their derivatives for alternative diesel fuels in CI engines: A review. *Renew. Sustain. Energy Rev.* **2011**, *15*, 131–149. [CrossRef]
13. Talebian-Kiakalaieh, A.; Amin, N.A.S.; Mazaheri, H. A review on novel processes of biodiesel production from waste cooking oil. *Appl. Energy* **2013**, *104*, 683–710. [CrossRef]
14. Singhabhandhu, A.; Tezuka, T. The waste-to-energy framework for integrated multi-waste utilization: Waste cooking oil, waste lubricating oil, and waste plastics. *Energy* **2010**, *35*, 2544–2551. [CrossRef]
15. Singhabhandhu, A.; Tezuka, T. Prospective framework for collection and exploitation of waste cooking oil as feedstock for energy conversion. *Energy* **2010**, *35*, 1839–1847. [CrossRef]
16. Salemdeeb, R.; zu Ermgassen, E.K.H.J.; Kim, M.H.; Balmford, A.; Al-Tabbaa, A. Environmental and health impacts of using food waste as animal feed: A comparative analysis of food waste management options. *J. Clean. Prod.* **2017**, *140*, 871–880. [CrossRef]
17. Tres, A.; Bou, R.; Guardiola, F.; Nuchi, C.D.; Magrinya, N.; Codony, R. Use of recovered frying oils in chicken and rabbit feeds: Effect on the fatty acid and tocol composition and on the oxidation levels of meat, liver and plasma. *Animal* **2013**, *7*, 505–517. [CrossRef] [PubMed]
18. Panadare, D.C.; Rathod, V.K. Applications of Waste Cooking Oil Other Than Biodiesel: A Review. *Iran. J. Chem. Eng.* **2015**, *12*, 55–76.
19. Sun, D.; Lu, T.; Xiao, F.; Zhu, X.; Sun, G. Formulation and aging resistance of modified bio-asphalt containing high percentage of waste cooking oil residues. *J. Clean. Prod.* **2017**, *161*, 1203–1214. [CrossRef]
20. Asli, H.; Ahmadinia, E.; Zargar, M.; Karim, M.R. Investigation on physical properties of waste cooking oil-Rejuvenated bitumen binder. *Constr. Build. Mater.* **2012**, *37*, 398–405. [CrossRef]
21. Worthington, M.J.H.; Kucera, R.L.; Albuquerque, I.S.; Gibson, C.T.; Sibley, A.; Slattery, A.D.; Campbell, J.A.; Alboaiji, S.F.K.; Muller, K.A.; Young, J.; et al. Laying Waste to Mercury: Inexpensive Sorbents Made from Sulfur and Recycled Cooking Oils. *Chem. Eur. J.* **2017**, *23*, 16219–16230. [CrossRef]
22. Lhuissier, M.; Couvert, A.; Amrane, A.; Kane, A.; Audi, J.-L. Characterization and selection of waste oils for the absorption and biodegradation of VOC of different hydrophobicities. *Chem. Eng. Res. Design* **2018**, *138*, 482–489. [CrossRef]
23. Global No. 1 Business Data Platform. Available online: https://www.statista.com/statistics/263937/vegetable-oils-global-consumption (accessed on 13 June 2020).
24. European Biomass Industry Association. Available online: https://www.eubia.org/cms/wiki-biomass/biomass-resources/challenges-related-to-biomass/used-cooking-oil-recycling/ (accessed on 3 July 2020).
25. Official Website of the European Union. Available online: https://eur-lex.europa.eu/legal-content/EN/TXT/?uri=CELEX%3A32014D0955 (accessed on 13 June 2020).
26. Official Website of the European Union. Available online: https://eur-lex.europa.eu/eli/dir/2008/98/oj (accessed on 13 June 2020).
27. Official Website of the European Union. Available online: https://eur-lex.europa.eu/legal-content/EN/TXT/?uri=CELEX:02000D0532-20150601 (accessed on 13 June 2020).
28. Official Website of the European Union. Available online: https://eur-lex.europa.eu/legal-content/EN/TXT/PDF/?uri=CELEX:32018L0851&from=EN (accessed on 13 June 2020).
29. Official Website of the European Union. Available online: https://eur-lex.europa.eu/eli/dir/2004/35/oj (accessed on 13 June 2020).
30. Official Website of the European Union. Available online: https://eur-lex.europa.eu/eli/reg/2011/142/oj (accessed on 13 June 2020).
31. Javier, R. Spain Only Collects Ten Percent of Used Cooking Oil and Europe Imports Millions of Litres to Produce Biodiesel. *Renew. Energy* **2019**. Available online: https://www.energias-renovables.com/biocarburantes/espana-solo-recoge-el-10-del-aceite-20191120 (accessed on 13 June 2020).
32. Official Website of the European Union. Available online: https://eur-lex.europa.eu/legal-content/EN/TXT/PDF/?uri=CELEX:32018L2001&from=EN (accessed on 13 June 2020).
33. ENERINVEST. Available online: https://www.enerinvest.es/noticias/general/el-24-de-diciembre-de-2018-entro-en-vigor-la-nueva-legislacion-europea-sobre-energias-renovables-eficienca-energetica-y-gobernanza-251 (accessed on 13 June 2020).

34. Instituto Para la Diversificación. Available online: https://www.idae.es/informacion-y-publicaciones/marco-legislativo-2030-el-paquete-de-invierno (accessed on 13 June 2020).
35. European Commission. Available online: https://ec.europa.eu/jrc/en/publication/raw-materials-demand-wind-and-solar-pv-technologies-transition-towards-decarbonised-energy-system (accessed on 4 July 2020).
36. Ministry for Ecological Transition and Demographic Challenge. Available online: https://www.miteco.gob.es/es/calidad-y-evaluacion-ambiental/temas/economia-circular/comision-europea/ (accessed on 13 June 2020).
37. Official Website of the European Union. Available online: https://eur-lex.europa.eu/resource.html?uri=cellar:ea0f9f73-9ab2-11ea-9d2d-01aa75ed71a1.0001.02/DOC_1&format=PDF (accessed on 13 June 2020).
38. Report 2018 CONOE. Available online: http://www.conoe.it/wp-content/uploads/2018/11/ANNUAL-REPORT-2018.pdf (accessed on 13 June 2020).
39. Higher Institute for Environmental Protection and Research. Available online: https://www.isprambiente.gov.it/it/garante_aia_ilva/normativa/normativa-ambientale/Dlgs_152_06_TestoUnicoAmbientale.pdf (accessed on 13 June 2020).
40. Chamber of Deputies. Available online: https://www.camera.it/parlam/leggi/deleghe/testi/10205dl.htm (accessed on 13 June 2020).
41. Novamont. Available online: https://www.novamont.com/matrol-bi (accessed on 13 June 2020).
42. Maire Tecnimont. Available online: https://www.mairetecnimont.com/en/media/press-releases/new-bio-lubricants-residual-fats-recovery-non-price-sensitive (accessed on 13 June 2020).
43. Official Website of the European Union. Available online: https://eur-lex.europa.eu/legal-content/EN/TXT/?uri=CELEX%3A32011R0459 (accessed on 7 July 2020).
44. Real Decreto 1085/2015. Available online: https://www.boe.es/buscar/doc.php?id=BOE-A-2015-13208 (accessed on 13 June 2020).
45. GEREGRAS. Available online: http://www.geregras.es/noticias/News/show/los-resultados-provi-3228 (accessed on 13 June 2020).
46. Real Decreto 1597/2011. Available online: https://www.boe.es/buscar/doc.php?id=BOE-A-2011-17465 (accessed on 13 June 2020).
47. Orden TEC/1420/2018. Available online: https://www.boe.es/diario_boe/txt.php?id=BOE-A-2018-18004 (accessed on 13 June 2020).
48. Real Decreto 235/2018. Available online: https://www.boe.es/diario_boe/txt.php?id=BOE-A-2018-5890 (accessed on 13 June 2020).
49. Instituto para la Diversificación. Available online: https://www.idae.es/informacion-y-publicaciones/plan-nacional-integrado-de-energia-y-clima-pniec-2021-2030 (accessed on 13 June 2020).
50. Ley 22/2011. Available online: https://www.boe.es/buscar/act.php?id=BOE-A-2011-13046 (accessed on 13 June 2020).
51. Specification about the Normative Related to Wastes and to SANDACH to the Treatment and Recycling of Animal by-Products. Ministerio de Agricultura y Pesca, Alimentación y Medio Ambiente (MAPAMA). Available online: https://www.mapa.gob.es/es/ganaderia/temas/sanidad-animal-higiene-ganadera/sandach/documentacion-interes/otros.aspx (accessed on 7 July 2020).
52. ReFood. Available online: https://www.refood.es/es/refoodes/biblioteca-certificados/ (accessed on 13 June 2020).
53. Ministry for Ecological Transition and Demographic Challenge. Available online: https://www.miteco.gob.es/es/calidad-y-evaluacion-ambiental/temas/prevencion-y-gestion-residuos/convenios-acuerdos-voluntarios/Convenio-voluntario-6-con-FEHR-GEREGRAS.aspx (accessed on 13 June 2020).
54. Analysis of the Current Development of Household UCO Collection Systems in the EU. GREENEA (2016). Available online: https://theicct.org/publications/analysis-current-development-household-uco-collection-systems-eu (accessed on 13 June 2020).
55. Press Release of the General Assembly 2019 GEREGRAS. Available online: http://www.geregras.es/noticias/News/show/geregras-se-reune-p-3201 (accessed on 7 July 2020).
56. GEREGRAS (2019). Available online: http://www.geregras.es/noticias/News/show/geregras-se-reune-p-3201 (accessed on 13 June 2020).
57. Euractiv. Available online: https://www.euractiv.com/section/agriculture-food/opinion/fraudulent-used-cooking-oil-biodiesel-bad-for-the-climate-and-a-blow-to-eu-farm-oilseed-and-plant-protein-sectors/ (accessed on 13 June 2020).

58. Vlahopoulou, G.; Petretto, G.L.; Garroni, S.; Piga, C.; Mannu, A. Variation of density and flash point in acid degummed waste cooking oil. *J. Food Proc. Pres.* **2018**, *42*, e13533. [CrossRef]
59. Mannu, A.; Vlahopoulou, G.; Urgeghe, P.; Ferro, M.; Del Caro, A.; Taras, A.; Garroni, S.; Rourke, J.P.; Cabizza, R.; Petretto, G.L. Variation of the chemical composition of waste cooking oils upon bentonite filtration. *Resources* **2019**, *8*, 108. [CrossRef]
60. Di Pietro, M.E.; Mannu, A.; Mele, A. NMR determination of free fatty acids in vegetable oils. *Processes* **2020**, *8*, 410. [CrossRef]

Article

Prickly Pear Seed Oil by Shelf-Grown Cactus Fruits: Waste or Maste?

Vassilios K. Karabagias, Ioannis K. Karabagias *, Ilias Gatzias and Anastasia V. Badeka

Laboratory of Food Chemistry, Department of Chemistry, University of Ioannina, 45110 Ioannina, Greece; vkarambagias@gmail.com (V.K.K.); iliasgr1985@yahoo.gr (I.G.); abadeka@uoi.gr (A.V.B.)
* Correspondence: ikaraba@cc.uoi.gr; Tel.: +30-697-828-686-6

Received: 24 December 2019; Accepted: 16 January 2020; Published: 21 January 2020

Abstract: The chemical composition and properties of seed oils have attracted researchers nowadays. By this meaning, the physicochemical and bioactivity profile of prickly pear seed oil (PPSO) (a product of prickly pear fruits waste) were investigated. Seeds of shelf-grown cactus fruits (*Opuntia ficus indica* L.) were subjected to analysis. Moisture content (gravimetric analysis), seed content (gravimetric analysis), oil yield (Soxhlet extraction/gravimetric analysis), volatile compounds (HS-SPME/GC-MS), fatty acids profile (GC-FID), *in vitro* antioxidant activity (DPPH assay), and total phenolic content (Folin-Cioacalteu assay) were determined. Results showed that prickly pear seeds had a moisture content of 6.0 ± 0.1 g/100 g, whereas the oil yield ranged between 5.4 ± 0.5 g/100 g. Furthermore, the PPSO had a rich aroma because of acids, alcohols, aldehydes, esters, hydrocarbons, ketones, and other compounds, with the major volatiles being 2-propenal, acetic acid, pentanal, 1-pentanol, hexanal, 2-hexenal, heptanal, 2-heptenal (Z), octanal, 2-octenal, nonanal, 2,4-decadienal (E,E), and trans-4,5-epoxy-(E)-2-decenal. Among the fatty acids, butyric, palmitic, stearic, and oleic acids were the dominant. Finally, the pure PPSO had a high *in vitro* antioxidant activity (84 ± 0.010%) and total phenolic content (551 ± 0.300 mg of gallic acid equivalents/L). PPSO may be then used as a beneficial by-product, in different food systems as a flavoring, antioxidant, and nutritional agent.

Keywords: PPSO; volatiles; FAs; antioxidant activity; phenolics

1. Introduction

The exploitation of the food waste and the use of new materials, nowadays, has become more than ever a potential demand for the humanity for the future welfare and existence. The by-products of processed fruits (i.e., peel and seeds) could comprise a new and effective source of oil and food given their nutrients including bio-flavonoids, proteins, minerals, fatty acids, etc., [1]. The fact that millions of pounds of seeds, originating from different fruits, are discarded every year with no strategic and programmed disposal, leads to environmental concern [2]. *Opuntia ficus-indica* (L.) Mill. or prickly pear, is a tropical or subtropical plant of the Cactaceae family and is commonly used for fruit production. Given that it is a good source of natural antioxidants, hence, it can be used in foods or nutritional supplements [1].

On the other hand, the seed oils of fruits are of great interest because these are edible oils (possessing a high degree of unsaturation) with antioxidant, antimicrobial, and biological activity [1–4]. Therefore, the oil from seeds can be potentially used by the food industry for the production of natural-based foods [5], with extended shelf-life [6,7]. More specifically, the oil from cactus pear seeds has been reported to have considerable amounts of unsaturated fatty acids [1], and antioxidant [8,9] or antimicrobial activity [10], as well as cardioprotective, anti-thrombotic, anti-inflammatory, anti-arrhythmic, hypolipidemic, and anti-hyperglycemic properties [11,12]. These properties are of interest for the pharmaceutical and food sector. However, the yield of prickly pear seed oil may vary among cultivars, crop environmental factors/geographical origin (i.e., light, temperature, rainfall, and type of soil nutrients), or methods and solvents used for the extraction [2,13]. Based on the aforementioned, the purpose of the present study was to determine the moisture content, seed content, oil yield, volatile compounds, free fatty acids profile, *in vitro* antioxidant activity, and total phenolic content of prickly pear seed oil, extracted from prickly pear fruits of the wild cultivar, grown in the region of Messinia (Peloponnese). Different procedures were followed including the use of either pure prickly pear seed oil (PPSO) or its methanolic extract for the characterization of seed oil antioxidant activity and total phenolic content.

It is worth mentioning, that this is the first report in the literature on the different physicochemical properties of PPSO obtained from Greek wild prickly pear cultivars. Present findings support the relevant literature, and may contribute to comparative studies dealing with the characterization and beneficial use of PPSO from different regions, concerning primarily the food or pharmaceutical industry.

2. Materials and Methods

2.1. Prickly Pear Fruits and Seeds

Prickly pear fruits from the shelf-grown cultivar (*Opuntia ficus indica* L.), locally termed as the "wild" cultivar from the region of Messinia (Peloponnese) were used in the study to estimate first the contribution of the seeds to the total fruit mass. Randomly chosen fruits of yellow to green color, were peeled and cut in four pieces. Then, these were left to lose all the water nutrients at room temperature for 24 h. The next day the seeds were removed from the fruit with a niger, and weighted in a Sartorius balance. For the isolation of prickly pear seed oil (PPSO), approximately 5 kg of prickly pear seeds (originating from 1 ton of seeds) were provided by a local agricultural cooperative in the region of Messinia during the harvesting season 2015. The monthly climatological summary for August 2017, during which fruits had the optimum maturity, included the consideration of average temperature (°C), wind speed (km/h), and rainfall (mm). The respective values were 28.1 °C, 4.8 (km/h), and 2.8 (mm). Data were provided by the National Observatory of Athens. The number of the independently obtained PPSOs was $n = 3$. Afterwards, the PPSOs were combined and subjected to analysis.

2.2. Chemicals and Reagents

The chemicals and reagents used in the study such as gallic acid (3,4,5-trihydrobenzoic acid), methanol (MeOH), acetate buffer ($CH_3COONa{\cdot}3H_2O$), Folin-Ciocalteu phenol reagent, sodium chloride (NaCl), potassium hydroxide(KOH), and sodium carbonate (Na_2CO_3), were purchased from Merck (Darmstadt, Germany). The stable free radical 2,2-diphenyl-1-picrylhydrazyl (DPPH) was purchased from Sigma-Aldrich (Darmstadt, Germany).

2.3. Determination of Moisture Content of PPSO

About 25 ± 1 g of prickly pear seeds were introduced in a metallic plate of low bas (Brieger, Sclieren, Switzerland) and dried at 105 °C in a Memmert (Memmert GmbH + Co, KG, Schwabach, Germany) drying oven until constant weight. Results were expressed as g/100 g.

2.4. PPSO Isolation Using Soxhlet Extraction

The PPSO was extracted using Soxhlet extraction. In particular, 25 ± 1 g of dried seeds were blended for 20 min (AVEC blender, Jumbo S.A., Athens, Greece) and placed in an extraction thimble of 30 × 100 mm size (Filtres Fioroni, Ingré, France) and introduced in the Soxhlet apparatus. The organic solvent used for the extraction was *n*-hexane (Merck, Darmstadt, Germany). The extraction was completed in 4–6 h. The *n*-hexane was evaporated under vacuum rotation (Büchi waterbath B-480, Büchi rotavapor R-114, Flawil, Switzerland). The prickly pear seed oil was then re-suspended in *n*-hexane, dried at 105 ± 1 °C for 2 h, and collected in vials. The drying process was carried out to ensure that no solvent residues were present. The extraction was carried out in triplicate ($n = 3$).

2.5. Determination of the Volatile Compounds of Prickly Pear Seed Oil (PPSO)

The PPSO volatile compounds were determined using headspace solid phase micro-extraction coupled to gas chromatography/mass spectrometry (HS-SPME/GC-MS). The divinyl benzene/carboxen/polydimethylsiloxane (DVB/CAR/PDMS) fiber 50/30 μm (Supelco, Bellefonte, PA, USA) was used for the extraction of PPSO volatile compounds. The PPSO samples (ca. 2 g), were placed in screw-cap vials of 15 mL volume, equipped with PTFE/silicone septa. For the headspace extraction, the vials were maintained at 45 °C in a water bath under stirring at 600 rpm. A magnetic stirrer of 10 mm diameter, coated with polytetrafluoroethylene (PTFE) (Semadeni, Ostermundigen–Bern, Switzerland) was placed inside the vials. The analysis conditions were: 30 min equilibration time, 15 min sampling time, 2 g sample mass, and 45 °C water bath temperature. Each PPSO sample was run in duplicate (n = 2).

2.6. GC/MS Instrumentation and Conditions of Analysis

The SHIMADZU GC-2010 Plus series gas chromatograph was used for the analysis of the volatile compounds of PPSO. A DB-5MS (cross linked 5% PH ME siloxane) capillary column (60 m × 250 μm i.d., × 1 μm film thickness) was used, with helium as the carrier gas. The total flow rate was 6.7 mL min^{-1}, whereas that of the column was 3.69 mL min^{-1} and the purge flow was 3.0 mL min^{-1}. The pressure during the analysis was 320 kPa. The injector temperature was 260 °C, respectively. For the SPME analysis, the oven temperature was held at 40 °C for 5 min, increased to 160 °C at 3 °C/min (0 min hold), and finally increased to 240 °C at 10 °C/min (17 min hold). The total program time was 70 min.

2.7. Identification of the Volatile Compounds of PPSO

The identification of compounds was achieved using the NIST 2011 (NIST-11) mass spectral library by considering the GC-MS spectra.

2.8. Determination of Free Fatty Acids (FAs) of PPSO

The analysis of FAs was carried out using gas chromatography coupled to flame ionization detector (GC-FID). The composition of FAs was determined after transesterification into fatty acid methyl esters (FAMEs). The PPSO was diluted in heptane (0.1 g in 2 mL) with 0.2 mL of 2 mol L^{-1} potassium hydroxide in methanol, in a test tube with a screw cap and was vigorously shaken, to obtain the methyl esters. The mixture was then centrifuged and the supernatant layer containing the methyl esters was used for the GC analysis. Before the analysis, the mixture was filtered using PTFE membrane filters of 13 mm pore size (Millex-FG, Merck, Darmstadt, Germany). The FAMEs were analyzed on a GC-FID chromatograph (model 6890 N, Agilent Technologies, Wilmington, DE, USA), equipped with a 30 m × 320 μm i.d.,Supelcowax column with a film thickness of 0.5 μm (Supelco, Sigma-Aldrich, Darmstadt, Germany). The carrier gas was helium, at a flow rate of 1.5 mL min^{-1}. The temperatures of the injector and detector were set at 250 and 260 °C, respectively. The initial oven temperature was 60 °C held for 5 min and increased from 60 to 250 °C at a rate of 7 °C min^{-1}. Finally, it was held at 250 °C for 2 min. The injection volume was 1 μL. A split ratio of 20:1 was used [14]. Each sample was analyzed in triplicate (n = 3).

2.9. Determination of In Vitro Antioxidant Activity (IVAA) and Total Phenolic Content (TPC) of PPSO

The antioxidant activity of either pure PPSO or its methanolic extracts was estimated *in vitro* using the DPPH assay [15]. A standard solution of DPPH (43 mg/L) was prepared by dissolving 0.0043 g of DPPH in 100 mL of methanol. For antioxidant activity test 1.8 mL DPPH plus 1 g (or 1 mL) of pure PPSO (or its methanolic extracts), plus 0.20 mL of the acetate buffer (pH = 6.9 ± 0.1) were mixed in a cuvette (final volume of 3 mL) and the absorbance of the reaction mixture was measured at 517 nm (Spectrophotometer Model UV-1280, Shimadzu, Kyoto, Japan). It is worth mentioning, that the extraction of PPSO with methanol was carried out after centrifugation of the sample for 20 min at 4000 rpm (model of centrifuge: Biofuge, primo R, Heraeus) (Kendro Laboratory Products, Osterode, Germany). The absorbance was measured every 60 min (regular time periods) until the value reached a plateau (steady state). The plateau was reached at 3 h. The IVAA was calculated using the following equation:

$$\%IVAA = ((A_0 - A_t)/A_0) \times 100 \quad (1)$$

where A_0 is the initial absorbance of the DPPH free radical standard solution and A_t is the absorbance of the remaining DPPH free radical, after reaction with the PPSO pure samples or its methanolic extracts at the plateau. For the estimation of the effective concentration that causes inhibition of the DPPH radical by 50% (EC_{50}) a graphical plot of the percent inhibition of the DPPH radical versus different proportions of either the pure PPSO or its methanolic extracts was prepared (y = ax + b), and the EC_{50} (g or mg/mL) was determined by the obtained linear equation by inserting the y-value (%DPPH inhibition) = 50. The blank sample (final volume of 3 mL) consisted of methanol plus buffer (2:1 *v/v*). All samples were filtered using Whatman polyethersulfone (PES) membrane filters (Fisher Scientific, Loughborough, Leicestershire, UK) with a pore size of 0.45 μm before absorbance measurements.

Similarly, the TPC of the pure PPSO or its methanolic extracts was determined using the Folin-Ciocalteu colorimetric method [16]. Briefly, in a 5 mL volumetric flask 0.20 g or 0.20 mL of the pure PPSO (or its methanolic extracts) followed by 2.50 mL of distilled water and 0.25 mL Folin-Ciocalteu reagent were added. After 3 min, 0.50 mL of saturated Na_2CO_3 (30% *w/v*) were also added into the mixture. Finally, the obtained solution was brought to 5 mL with distilled water. This solution was left for 2 h in the dark at room temperature and the absorbance was measured at 760 nm, after filtration with Whatman PES membrane filters, in the aforementioned UV/VIS spectrophotometer. A calibration curve using gallic acid (GA) was prepared between 0–780 mg/L:

$$y = 0.000700x + 0.0390,\ R^2 = 0.9695, \quad (2)$$

The TPC was expressed as mg of gallic acid equivalents (mg GAE/mL). For the IVAA and TPC determinations, each sample was analyzed in triplicate ($n = 3$).

2.10. Statistical Analysis

Correlations were obtained using Pearson's bivariate correlation coefficient (r), at the confidence level of $p < 0.05$. The t-test was applied at the confidence level of $p < 0.05$ for the comparison of average values. Statistical analysis was done using the SPSS (Statistical Package for the Social Sciences) version 20.0 statistics software (IBM Corp., Armonk, NY, USA, 2011). The average (±standard deviation) values of bioactivity parameter analyses were estimated using the Microsoft Office Excel spread sheets for Windows 2007 (Microsoft Corp., Redmond, WA, USA).

3. Results and Discussion

3.1. Contribution of Seeds to the Total Fruit Mass

The peeled fruits had an average weight of 47.3 g (ca. 47 g). The weight of the seeds was 3.82 g (ca. 3.8 g). Therefore, the seeds contributed by: 3.82/47.3 × 100 = 7.44%, to the total peeled fruit mass. There is scarce data available in the literature concerning the seed content of prickly pear fruits grown in Greece. Data on seed content of prickly pear fruits may involve the fruit processor in terms of the exploitation of seed extracts for the preparation of nutritional foodstuff. It has been reported in the literature that seeds of prickly pear fruits are a good source of protein (ca. 11.8%) [16].

3.2. Moisture Content of Prickly Pear Fruits and Seeds

The moisture content of prickly pear fruits was 67.3 ± 2.29 g/100 g. In a previous study dealing with prickly pear fruits from Morocco the reported moisture content values were significantly ($p < 0.05$) higher, in the range of 89.1 ± 8.09 to 91.2 ± 9.23 g/100 g [17]. Differences in the moisture content of the aforementioned prickly pear fruits indicate the impact of the geographical origin on fruit composition, and especially, the maturity level of fruit. The higher the moisture content the less is the condensation of sugars, and therefore, different will be the maturity level of fruit. In addition, different moisture content may probably give information about the water absorption capacity of the cultivar and the soil conditions. On the other hand, prickly pear seeds had a significantly ($p < 0.05$) lower moisture content than the fruit, in the range of 6.0 ± 0.1 g/100 g.

3.3. Volatile Compounds of PPSO

Two hundred and twenty one volatile compounds were identified in PPSO using the GC-MS spectra as shown in Supplementary Table S1. The volatile compounds could be grouped in acids (2.70%), alcohols (9.13%), aldehydes (62.72%), esters (2.82%), hydrocarbons (5.06%), ketones (4.38%), and other compounds (12.71%). It is characteristic that substantial differences ($p < 0.05$) were recorded among the classes of volatile compounds using t-test analysis (Figure 1). The most dominant volatile compounds were aldehydes. A typical gas chromatogram indicating the major volatile compounds of PPSO is shown in Figure 2. Among aldehydes, 2-propenal, pentanal, hexanal, 2-hexenal, heptanal, 2-heptenal, (Z), octanal, 2-octenal, nonanal, 2,4-decadienal (E,E), and trans-4,5-epoxy-(E)-2-decenal recorded the higher proportions (Table S1). There is limited data available in the literature, regarding the volatile composition of PPSO. Zito et al. [10] reported that the PPSO obtained from the Sanguigna cultivar grown in Sicily, containedmainly hydrocarbons (38.5%), fatty acids, and derivatives (31.9%) and terpenes (12.4%), whereas the PPSO obtained from the Surfarina cultivar grown in the same region, contained the highest amounts of fatty acids and derivatives (68.9%), followed by terpenes (10.9%). In the present study, the volatile compound 2-propenal (or acrolein) is the simplest unsaturated aldehyde and has a piercing and acrid smell. For example, when cooking oil is heated to its smoke point, a burnt fat odor is caused by glycerol in the burning fat, breaking down then into acrolein [18]. The alkyl aldehydes pentanal, hexanal, heptanal, octanal, and nonanal possess a fruit-like odor with diverse senses where their concentration differs. In particular, it has been reported that pentanal has a fermented, bready, fruity, nutty, and berry smell.Similarly, hexanal contributes to a hay-like "off-note" flavor in green peas [19]. Heptanal has a strong fruity odor and naturally occurs in the essential oils of ylang-ylang (*Cananga odorata*), clary sage (*Salvia sclarea*), lemon (*Citrus x limon*), bitter orange (*Citrus x aurantium*), rose (Rosa), and hyacinth (*Hyacinthus*) [20].

The compound (E,E)-2,4-decadienal is a volatile substance found in butter, cooked beef, fish, potato chips, roasted peanut, buckwheat, and wheat bread crumb. The smell intensity is related to its concentration. For example, it smells of deep fat flavor, characteristic of chicken aroma (at 10 ppm), whereas at lower concentration, it has the odor of citrus, orange or grapefruit [21]. Likewise, trans-4,5-epoxy-(E)-2-decenal is an oxygenated α,β-unsaturated aldehyde, and can be formed during the baking of fats that contain linoleic acid. The acids 13-hydroperoxy-9,11-octadecadienoic and 9-hydroperoxy-10,12-octadecadienoic are the intermediates in the biochemical process [22]. Aldehyde is also formed in cooked beef when it remains in the refrigerator for a long storage time, contributing to a fusty odor [23,24]. It is also an important part of the smell of raw and cooked mutton [22]. Apart from the aldehydes, acetic acid was identified in considerable proportions. Acetic acid is the second simplest carboxylic acid with a pungent, sour, and overripe fruit-like odor. It is the product of ethanol oxidation or fermentation by acetic acid bacteria (*Acetobacter* and *Clostridium acetobutylicum*). These bacteria are commonly found in foodstuff, water, and soil, and acetic acid is produced naturally as fruits and other foods spoil. Acetic acid is traditionally considered as a mild antibacterial agent [25]. Finally, 1-pentanol has been reported to possess a fermented-like, yeasty, bready, and fusel odor. Numerous of the volatile compounds of PPSO, such as pentanal, hexanal, heptanal, octanal, nonanal, (E)-2-octenal, decanal, benzaldehyde, acetone, 1-pentanol, 1-hexanol, 2-pentylfuran, tridecane, acetone, 6-methyl-5-hepten-2-one, 3-octen-2-one, and others, have been previously reported in the seeds of Canadian faba bean from different genotypes (*Vicia faba* L.) [26]. What is also remarkable, is that numerous of the volatile compounds of PPSO were identified previously in prickly pear juice prepared by the wild cactus fruits [27]. Based on the aforementioned, it is quite obvious that the volatile senses of PPSO are diverse, indicating thus, its potential multi-use in different materials and food matrices.

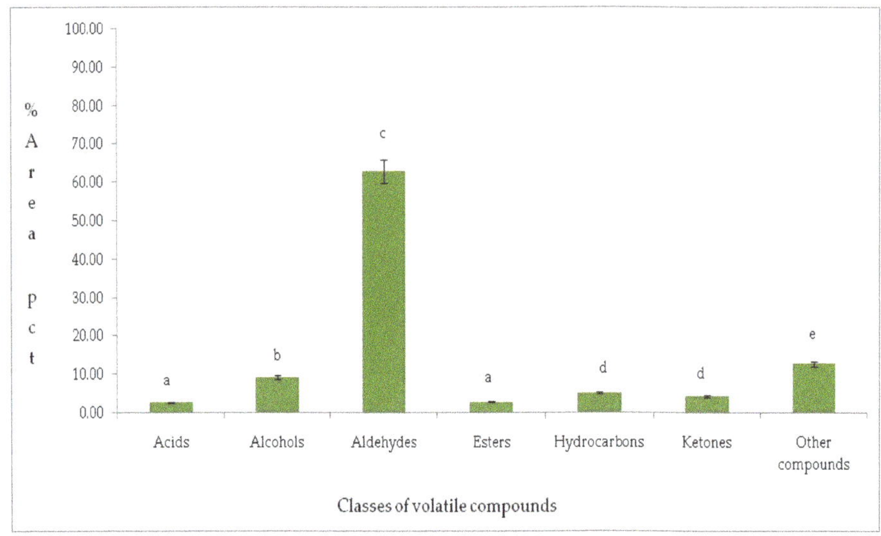

Figure 1. Proportions of classes of volatile compounds identified in prickly pear seed oil (PPSO). Different letters in each bar indicate statistically significant differences ($p < 0.05$).

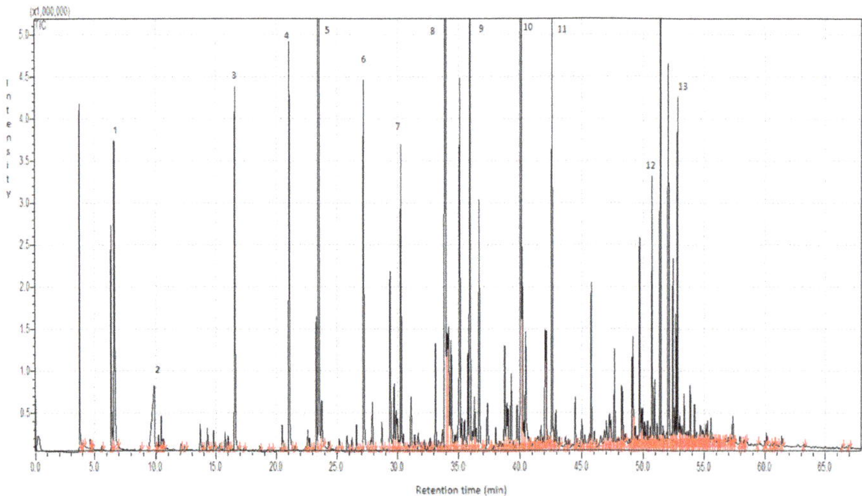

Figure 2. A typical gas-chromatogram of PPSO indicating with numbers the most dominant volatile compounds. 1: 2-propenal, 2: acetic acid, 3: pentanal, 4: 1-pentanol, 5: hexanal, 6: 2-hexenal, 7: heptanal, 8: 2-heptenal (Z), 9: octanal, 10: 2-octenal, 11: nonanal, 12: 2,4-decadienal (E, E), 13: *trans*-4,5-epoxy-(E)-2-decenal.

3.4. Oil Yield and Fatty Acid Profile of PPSO

The oil yield based on the grounding process with the household blender was 5.4 ± 0.5% (g/100 g). For example, the cold press method resulted in a significantly lower oil yield compared to maceration-percolation method in the study of Regalado-Renteria et al. [4], concerning the isolation of PPSO from eight different prickly pear fruit varieties. The cold press method resulted in oil yield ranging from 0.51 ± 0.01 to 6.1 ± 0.6 g/100 g, whereas that of the maceration-percolation method ranged between 6.2 ± 0.3 to 16 ± 0.48 g/100 g. At this point it should be stressed that the oil yield of seeds depends primarily on the prickly pear variety and second on the process of isolation that is followed. In the present study, the use of *n*-hexane for the extraction of prickly pear seed oil resulted in comparable results, especially with the oil yield obtained through the cold press method as reported in the study of Regalado-Renteria et al. [4]. In addition, no hexane residues were observed during the HS-SPME/GC/MS analysis, indicating the absence of any contamination. In the study of Ramírez-Moreno et al. [2] the oil extraction with hexane was higher for both fruit varieties of *Opuntia* (green and red cultivar) (11.83% and 6.89%, respectively), compared to ethanol or ethyl acetate. The use of maceration-percolation method, however, cannot be considered as the most effective procedure for the increase in seed oil yield, given the fact that varietal differentiation is the dominant parameter for the isolation of PPSO. Indeed, in the study of Morales et al. [3] the oil yield of *Xoconostle* seeds in Mexico ranged between 2.45 and 3.52%. In another study concerning the PPSO obtained from Algerian prickly pear cultivars (*Opuntia ficus indica*) the PPSO yield ranged between 7.3–9.3% [1]. The oil yield in Tynisian and Turkish PPSOs ranged between 9.88–11.75% and 5.00–14.4%, respectively [9,28–30], whereas the oil yield in prickly pear seeds of different South African varieties ranged between 2.24–5.69% [13]. The next step was to characterize the fatty acid profile of PPSO. The most dominant fatty acids were butyric acid (C4:0), palmitic acid (C16:0), stearic acid (C18:0), and oleic acid (C18:1) (Figure 3).

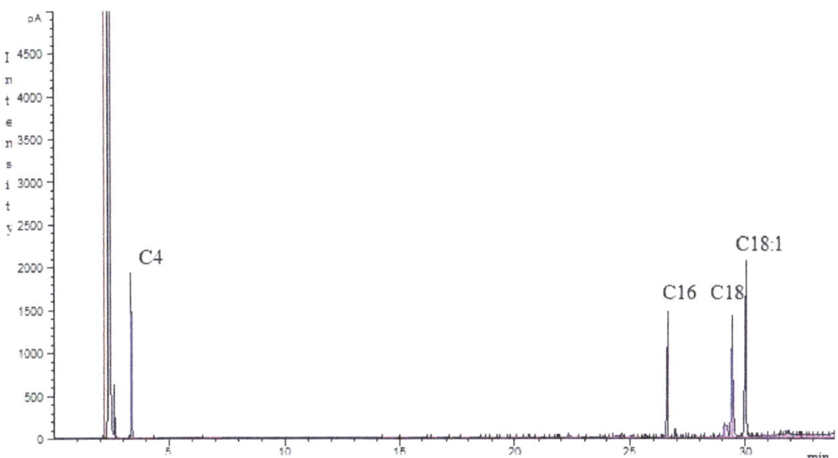

Figure 3. A typical gas chromatography-flame ionization detector (GC-FID) chromatogram indicating the fatty acids of PPSO.

The respective amounts were 0.48 ± 0.14%, 0.62 ± 0.02%, 0.68 ± 0.06%, and 1.56 ± 0.84%. The higher proportions were found for oleic acid. Therefore, PPSO could comprise an alternative source of oleic acid, among the other fatty acids. Previous results are in accordance with those of Regalado-Renteria et al. [4] in a study concerning the fatty acid content of PPSO obtained from different Mexican prickly pear fruit cultivars. South African PPSO showed higher proportions of palmitic, stearic, and oleic acids [13]. Italian PPSO obtained from the *Opuntia ficus indica*, Sanguigna and Surfarina varieties grown in Sicily, also showed higher proportions of palmitic, stearic, and oleic acids [31]. The same trend was also reported for Moroccan PPSO obtained from *Opuntia ficus indica* and *Opuntia dillenii* prickly pear fruits [32]. However, butyric acid was not reported in the above studies [13,31,32]. Butyric acid in combination with the (E,E)-2,4-decadienal identified in the volatile fraction, enhance further, the butter-like flavor of PPSO.

3.5. IVAA and TPC of PPSO

The DPPH assay has been used to predict the oxidative stability of edible oils [33]. The IVAA of the pure PPSO (initial amount of 0.2286–0.23 g) which was directly inserted in the cuvette containing 1.8 mL of DPPH plus 1 mL of the acetate buffer was 84 ± 0.010%, whereas that of the PPSO methanolic extract (mother solution of 0.5086 g PPSO/50 mL MeOH) was 54.21 ± 0.01% and that of the same amount of PPSO (~0.23 g) in 50 mL of methanol was 49.04 ± 0.02%. For the estimation of $EC5_0$ different proportions of either the pure PPSO or its methanolic extract were used to obtain a graph of DPPH inhibition versus amounts of pure PPSO or its methanolic extracts. In particular, for the pure PPSO the amounts used were: 0.2286 g, 0.1185 g, 0.056 g, 0.0331 g, and 0.0120 g. The obtained linear curve is shown in Figure 4a. The obtained EC_{50} value was estimated from the graph and was 0.096–0.1 g. In the case of PPSO methanolic extract, the amounts used were: 10.17 mg/mL, 5.09 mg/mL, 2.54 mg/mL, 1.02 mg/mL, and 0.51 mg/mL. Similarly, the EC_{50} value was 7.44 mg/mL (Figure 4b). As it can be observed from Figure 4a,b differences in the linearity (and intercept/slope) of the curves were obtained, depending on the use of either pure PPSO or its methanolic extract. In that sense, two major issues must be addressed: (i) the DPPH inhibition is dependent of the amount of the antioxidant used and (ii) the extraction of PPSO with methanol reduced substantially its IVAA compared to the direct use of pure PPSO. In a previous study dealing with the *in vitro* antioxidant capacity of Algerian prickly pear seeds [34], the authors reported that the best results of antioxidant capacity were obtained when the seeds were extracted with 75% acetone (among ethanol, methanol, and water 50%, *v*/*v*) using 0.2 g/10 mL of the extract. The obtained *in vitro* antioxidant capacity was 95%, in general agreement, with present results concerning the direct use of PPSO for the estimation of antioxidant activity. Prickly pear seed oil of Sicilian cultivars (Sanguigna and Surfarina) [29] obtained with Soxhlet extraction, showed the highest inhibitory concentration (IC_{50}) against the DPPH free radical, in agreement with present results. In addition, present results are in accordance with those of Ramírez-Moreno et al. [2], who reported a considerable antioxidant activity concerning the seed oil obtained from two Mexican prickly pear cultivars.

The same trend was also observed in the TPC of pure PPSO and those of its methanolic extracts. More specifically, the TPC of pure PPSO was 551 ± 0.300, whereas that of the methanolic extracts was significantly lower, ranging between 93.3 ± 0.140 mg/L. There was a perfect Pearson's correlation ($r = 1.000$, $p = 0.01$) between the IVAA and TPC content of pure and methanolic extracts of PPSO, indicating again the impact of solvent extraction. Even though solvent extraction is usually used for the isolation of antioxidants, the efficiency of the extraction depends on the selected solvent for the complete isolation/extraction of different antioxidant compounds with varying polarity [35]. The use of pure PPSO resulted in a much higher TPC compared to that of the methanolic extract. This is owed to the nature of the Folin-Ciocalteu assay given that it measures all the reducing agents (phytochemicals, fatty acids, minerals, etc.) being present in the pure PPSO, and not only the extracted antioxidants.

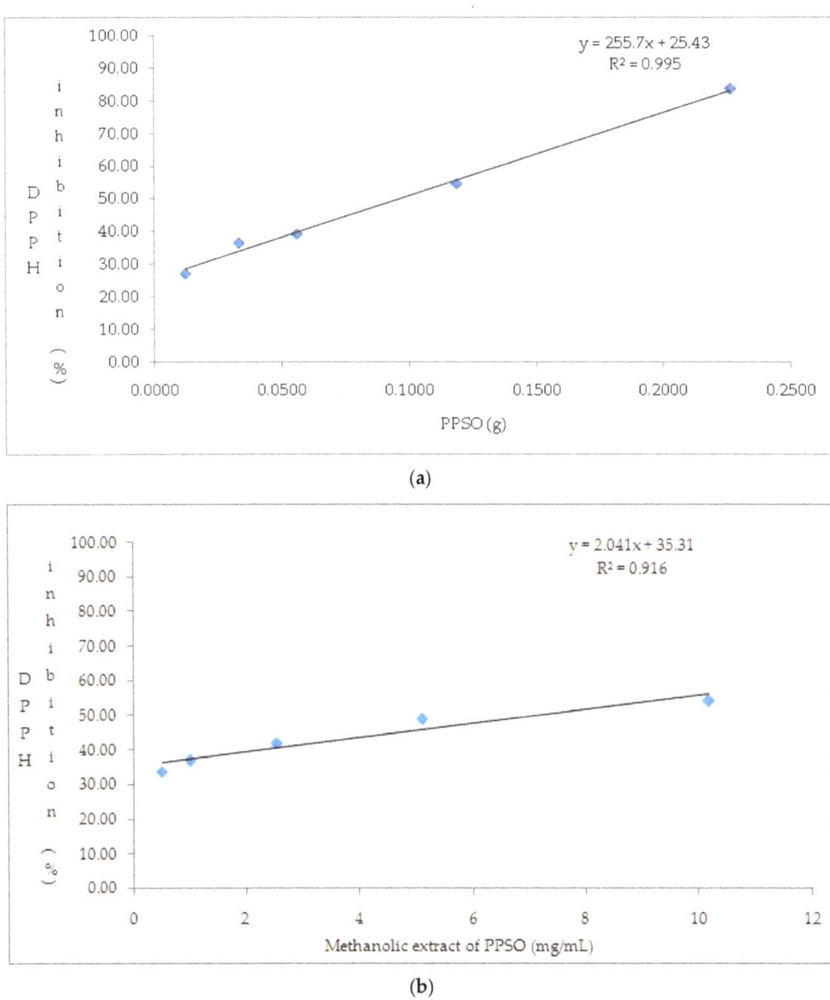

Figure 4. (**a**) 2,2-Diphenyl-1-picrylhydrazyl (DPPH) inhibition (%) of pure PPSO of different mass (g); (**b**) DPPH inhibition (%) of methanolic extract of PPSO of different concentration (mg/mL).

4. Conclusions

Results of the present study showed that prickly pear seed oil (PPSO) is a matrix of a rich aroma, considerable proportions of fatty acids, and high *in vitro* antioxidant activity in relation to the total phenolic content. What is worth mentioning, is that the PPSO obtained from the wild cultivars grown in Messinia (Peloponnese), showed the highest proportions of oleic acid, and considerable amounts of butyric acid, the latter reported for the first time in the relevant literature [13,31–33]. In addition, the rich fraction of over 200 volatile compounds of different class, gives a regional identity to the product. Apart from the cultivar impact [10], it should also be noted, that the chemical composition of PPSO may be affected by the harvesting season concerning the general weather conditions in a specific region [31]. Considering that the seeds of prickly pear are ignored at a global level during fruit consumption, and the PPSO is primarily used in cosmetics, the present study aims to reconsider the former use by proposing its application in food systems, either as flavoring and nutritious matrix, or as antimicrobial and antioxidant agent, in accordance with the cited literature. Therefore, the question in the title of this study has an answer: PPSO is a "maste." Future work with direct applications in different food systems will approve further the present findings.

Supplementary Materials: The following are available online at http://www.mdpi.com/2227-9717/8/2/132/s1. Table S1: Volatile compounds of PPSO tentatively identified using HS-SPME/GC-MS and NIST MS 11 mass spectral library.

Author Contributions: Conceptualization, V.K.K. and I.K.K.; methodology, I.K.K. and I.G.; software, A.V.B.; validation, I.K.K. and A.V.B.; formal analysis, V.K.K., I.K.K., I.G.; investigation, V.K.K. and I.K.K.; resources, A.V.B., I.K.K., V.K.K.; data curation, V.K.K., I.G., I.K.K.; writing—original draft preparation, I.K.K.; writing—review and editing, I.K.K.; visualization, V.K.K., I.K.K.; supervision, I.K.K., A.V.B.; project administration, I.K.K. All authors have read and agreed to the published version of the manuscript.

Funding: This research (2015-2017) received no external funding.

Acknowledgments: The authors are grateful to Maria Tasioula-Margari for the access she provided to the GC-MS instrument and to Alexandros Lolis for his technical assistance.

Conflicts of Interest: The authors declare no conflict of interest.

References

1. Chougui, N.; Tamendjari, A.; Hamidj, W.; Hallal, S.; Barras, A.; Richard, T.; Larbat, R. Oil composition and characterization of phenolic compounds of *Opuntia ficus-indica* seeds. *Food Chem.* **2013**, *139*, 796–803. [CrossRef]
2. Ramírez-Moreno, E.; Cariño-Cortés, R.; Cruz-Cansino, N.D.S.; Delgado-Olivares, L.; Ariza-Ortega, J.A.; Vanessa YelinaMontañez-Izquierdo, V.Y.; Hernández-Herrero, M.M.; Tomás Filardo-Kerstupp, T. Antioxidant and antimicrobial properties of cactus pear (*Opuntia*) seed oils. *J. Food Qual.* **2017**, *2017*, 1–8. [CrossRef]
3. Morales, P.; Ramírez-Moreno, E.; Sánchez-Mata, M.C.; Carvalho, A.M.; Ferreira, I.C. Nutritional and antioxidant properties of pulp and seeds of two xoconostle cultivars (*Opuntia joconostle* F.A.C. Weber ex Diguet and *Opuntia matudae* Scheinvar) of high consumption in Mexico. *Food Res. Int.* **2012**, *46*, 279–285. [CrossRef]
4. Regalado-Rentería, E.; Aguirre-Rivera, J.R.; González-Chávez, M.M.; Sánchez-Sánchez, R.; Martínez-Gutiérrez, F.; Juárez-Flores, B.I. Assessment of extraction methods and biological value of seed oil from eight variants of prickly pear fruit (*Opuntia* spp.). *Waste Biomass Valoriz.* **2018**, *9*, 1–9. [CrossRef]
5. Carović-Stanko, K.; Orlić, S.; Politeo, O.; Strikić, F.; Kolak, I.; Milos, M.; Satovic, Z. Composition and antibacterial activities of essential oils of seven *Ocimum* taxa. *Food Chem.* **2010**, *119*, 196–201. [CrossRef]
6. Tajkarimi, M.M.; Ibrahim, S.A.; Cliver, D.O. Antimicrobial herb and spice compounds in food. *Food Control* **2010**, *21*, 1199–1218. [CrossRef]
7. Solórzano-Santos, F.; Miranda-Novales, M.G. Essential oils from aromatic herbs as antimicrobial agents. *Curr. Opin. Biotechnol.* **2012**, *23*, 136–141. [CrossRef]
8. Liu, W.; Fu, Y.-J.; Zu, Y.-G.; Tong, M.-H.; Wu, N.; Liu, X.-L.; Zhang, S. Supercritical carbon dioxide extraction of seed oil from *Opuntia dillenii* Haw. and its antioxidant activity. *Food Chem.* **2009**, *114*, 334–339.
9. Matthäus, B.; Özcan, M.M. Habitat effects on yield, fatty acid composition and tocopherol contents of prickly pear (*Opuntia ficus-indica* L.) seed oils. *Sci. Hortic.* **2011**, *131*, 95–98. [CrossRef]
10. Zito, P.; Sajeva, M.; Bruno, M.; Rosselli, S.; Maggio, A.M.; Senatore, F. Essential oils composition of two Sicilian cultivars of *Opuntia ficus-indica* (L.) Mill. (Cactaceae) fruits (prickly pear). *Nat. Prod. Res.* **2013**, *27*, 1305–1314. [CrossRef] [PubMed]
11. Mobraten, K.; Haug, T.M.; Kleiveland, C.R.; Lea, T. Omega-3 and omega-6 PUFAs induce the same GPR120-mediated signalling events, but with different kinetics and intensity in Caco-2 cells. *Lipids Heal. Dis.* **2013**, *12*, 101. [CrossRef] [PubMed]
12. Berraaouan, A.; Ziyyat, A.; Mekhfi, H.; Legssyer, A.; Sindic, M.; Aziz, M.; Bnouham, M. Evaluation of antidiabetic properties of cactus pear seed oil in rats. *Pharm. Boil.* **2014**, *52*, 1286–1290. [CrossRef] [PubMed]
13. Labuschagne, M.; Hugo, A. Oil content and fatty acid composition of cactus pear seed compared with cotton and grape seed. *J. Food Biochem.* **2010**, *34*, 93–100. [CrossRef]
14. Gatzias, I.; Karabagias, I.; Kontakos, S.; Kontominas, M.; Badeka, A. Characterization and differentiation of sheep's milk from Greek breeds based on physicochemical parameters, fatty acid composition and volatile profile. *J. Sci. Food Agric.* **2018**, *98*, 3935–3942. [CrossRef] [PubMed]

15. Karabagias, V.K.; Karabagias, I.K.; Gatzias, I.; Riganakos, K.A. Characterization of prickly pear juice by means of shelf life, sensory notes, physicochemical parameters and bio-functional properties. *J. Food Sci. Technol.* **2019**, *56*, 3646–3659. [CrossRef] [PubMed]
16. El Kossori, R.L.; Villaume, C.; El Boustani, E.; Sauvaire, Y.; Méjean, L. Composition of pulp, skin and seeds of prickly pears fruit (*Opuntia ficus indica* sp.). *Plant Foods Hum. Nutr.* **1998**, *52*, 263–270. [CrossRef] [PubMed]
17. Dehbi, F.; Hasib, A.; Ouatmane, A.; Elbatal, H.; Jaouad, A. Physicochemical characteristics of Moroccan prickly pear juice (*Opuntia ficus indica* L.). *Int. J.Emerg. Technol. Adv. Eng.* **2014**, *4*, 300–306.
18. Martin, A.; Armbruster, U.; Atia, H. Recent developments in dehydration of glycerol toward acrolein over heteropolyacids. *Eur. J. Lipid Sci. Technol.* **2011**, *114*, 10–23. [CrossRef]
19. Roland, W.S.U.; Pouvreau, L.; Curran, J.; van de Velde, F.; de Kok, P.M.T. Flavor aspects of pulse ingredients. *Cereal Chem.* **2016**, *94*, 58–65. [CrossRef]
20. Burdock, G.A. *Fenaroli's Handbook of Flavor Ingredients*, 5th ed.; CRC Press: Boca Raton, FL, USA, 2005.
21. Vermeulen, N.; Czerny, M.; Gänzle, M.G.; Schieberle, P.; Vogel, R.F. Reduction of (E)-2-nonenal and (E,E)-2,4-decadienal during sourdough fermentation. *J. Cereal Sci.* **2007**, *45*, 78–87. [CrossRef]
22. Gassenmeier, K.; Schieberle, P. Formation of the intense flavor compound trans-4,5-epoxy-(E)-2-decenal in thermally treated fats. *J. Am. Oil Chem. Soc.* **1994**, *71*, 1315–1319. [CrossRef]
23. Konopka, U.C.; Grosch, W. Potent odorants causing the warmed-over flavour in boiled beef. *Eur. Food Res. Technol.* **1991**, *193*, 123–125. [CrossRef]
24. Rota, V.; Schieberle, P. Changes in key odorants of sheep meat induced by cooking. In *Food Lipids;* Chapter 6; ACS Symposium Series; American Chemical Society: Washington, DC, USA, 2005; Volume 920, pp. 73–83.
25. Buckingham, J. *Dictionary of Organic Compounds 1*, 6th ed.; Chapman & Hall: London, UK, 1996.
26. Oomah, B.D.; Razafindrainibe, M.; Drover, J.C. Headspace volatile components of Canadian grown low-tannin faba bean (*Vicia faba* L.) genotypes. *J. Sci. Food Agric.* **2014**, *94*, 473–481. [CrossRef]
27. Karabagias, V.K.; Karabagias, I.K.; Louppis, A.; Badeka, A.; Kontominas, M.G.; Papastephanou, C. Valorization of prickly pear juice geographical origin based on mineral and volatile compound contents using LDA. *Foods* **2019**, *8*, 123. [CrossRef]
28. El Mannoubi, I.; Barrek, S.; Skanji, T.; Casabianca, H.; Zarrouk, H. Characterization of *Opuntia ficus indica* seed oil from Tu;nisia. *Chem. Nat. Compd.* **2009**, *45*, 616–620. [CrossRef]
29. Ennouri, M.; Evelyne, B.; Laurence, M.; Hamadi, A. Fatty acid composition and rheological behavior of prickly pear seed oils. *Food Chem.* **2005**, *93*, 431–437. [CrossRef]
30. Ramadan, M.F.; Mörsel, J.-T.; Hassanien, M.F.R. Oil cactus pear (*Opuntia ficus-indica* L.). *Food Chem.* **2003**, *82*, 339–345. [CrossRef]
31. Loizzo, M.R.; Bruno, M.; Balzano, M.; Giardinieri, A.; Pacetti, D.; Frega, N.G.; Vincenzo, S.; Leporini, M.; Tundis, R. Comparative chemical composition and bioactivity of *Opuntia ficus-indica* Sanguigna and Surfarina seed oils obtained by traditional and ultrasound-assisted extraction procedures. *Eur. J. Lipid Sci. Technol.* **2019**, *121*, 1800283. [CrossRef]
32. Ghazi, Z.; Ramdani, M.; Fauconnier, M.L.; El Mahi, B.; Cheikh, R. Fatty acids sterols and vitamin E composition of seed oil of *Opuntia ficus indica* and *Opuntia dillenii* from Morocco. *J. Mater. Environ. Sci.* **2013**, *4*, 967–972.
33. Fazio, A.; Plastina, P.; Meijerink, J.; Witkamp, R.F.; Gabriele, B. Comparative analyses of seeds of wild fruits of *Rubus* and *Sambucus* species from Southern Italy: Fatty acid composition of the oil, total phenolic content, antioxidant and anti-inflammatory properties of the methanolic extracts. *Food Chem.* **2013**, *140*, 817–824. [CrossRef]
34. Chaalal, M.; Touati, N.; Louaileche, H. Extraction of phenolic compounds and *in vitro* antioxidant capacity of prickly pear seeds. *Acta Bot. Gallica* **2012**, *159*, 467–475. [CrossRef]
35. Soong, Y.-Y.; Barlow, P.J. Antioxidant activity and phenolic content of selected fruit seeds. *Food Chem.* **2004**, *88*, 411–417. [CrossRef]

Article

Interception Characteristics and Pollution Mechanism of the Filter Medium in Polymer-Flooding Produced Water Filtration Process

Xingwang Wang *, Xiaoxuan Xu, Wei Dang, Zhiwei Tang, Changchao Hu and Bei Wei

Petroleum Exploration and Production Research Institute, SINOPEC, Beijing 100083, China; xuxx.syky@sinopec.com (X.X.); dangwei.syky@sinopec.com (W.D.); tangzw.syky@sinopec.com (Z.T.); hucc.syky@sinopec.com (C.H.); weibei.syky@sinopec.com (B.W.)

* Correspondence: xingwang.syky@sinopec.com; Tel.: +86-010-8231-1523

Received: 16 October 2019; Accepted: 29 November 2019; Published: 5 December 2019

Abstract: Polymer flooding enhances oil recovery, but during the application of this technology, it also creates a large amount of polymer-contained produced water that poses a threat to the environment. The current processing is mainly focused on being able to meet the re-injection requirements. However, many processes face the challenges of purifying effect, facilities pollution, and economical justification in the field practice. In the present work, to fully understand the structure and principle of the oil field filter tank, and based on geometric similarity and similar flow, a set of self-designed filtration simulation devices is used to study the treatment of polymer-contained produced water in order to facilitate the satisfaction of the water injection requirements for medium- and low-permeability reservoirs. The results show that, due to the existence of polymers in oil field produced water, a stable colloidal system is formed on the surface of the filter medium, which reduces the adsorption of oil droplets and suspended solids by the filter medium. The existence of the polymers also increases the viscosity of water, promotes the emulsification of oil pollution, and increases the difficulty of filtration and separation. As filtration progresses, the adsorption of the polymers by the filter medium bed reaches saturation, and the polymers and oil pollution contents in the filtered water increase gradually. The concentration and particle size of the suspended solids eventually exceed the permissible standards for filtered water quality; this is mainly due to the unreasonable size of the particle in relation to the filter medium gradation and the competitive adsorption between the polymers and the suspended solids on the surface of the filter medium. The oil concentration of the filtered water also exceeds the allowable standards and results from the polymers replace the oil droplets in the pores and on the surfaces of the filter medium. Moreover, the suspended particles of the biomass, composed of dead bacteria, hyphae, and spores, have strong attachment and carrying ability with respect to oil droplets, which cause the suspended solids in the filtered water to exceed the permissible standards and oil droplets to be retained in the filtered effluent at the same time.

Keywords: polymer-flooding produced water; filtration; interception characteristics; filter medium pollution

---

1. Introduction

Polymer flooding is the leading technology for tertiary oil recovery in oil fields. This technology benefits the oil well and increases production. At the same time, however, residual polyacrylamide is a byproduct of this method of oil recovery, and a large amount of polymer-contained produced water is also produced [1,2]. This kind of produced water can be processed to meet the standards of the water quality index for oil content and suspended solids of less than 20 mg/L by undergoing traditional "two-stage" treatment (two-stage sedimentation and one-stage filtration). After treatment, the refined effluent can then be injected back into a high-permeability reservoir to meet the needs of oil field development and production [3–6]. However, the current trend in the oil industry is to further diversify the development of medium- and low-permeability reservoirs on the basis of high-permeability reservoir development. With the rise of comprehensive water cuts in oil fields and the application of polymer flooding industrialization, the scale of polymer-contained produced water is increasing [7]. Taking Daqing Oilfield as an example, 9×10^7 t of polymer-flooding produced water is produced every year under the current output [8]. It may appear that the scale of water source is enough to deal with the problem of valuable clear water resources and shortage of deep produced water source in oil field; however, the polymer-flooding produced water has the characteristics of high polymer concentration and suspended solids content. The polymer concentration is between 10 and 1000 mg/L, and the suspended solids content can reach up to 300 mg/L. In order to meet the re-injection requirements of treated produced water, the treatment is required to meet the standards. However, the presence of polymers in the polymer-flooding produced water increases the degree of emulsification, posing a challenge to the entire produced water treatment process. Furthermore, the advanced treatment technology of polymer-contained produced water is not yet mature, which still leads to the problem of water imbalance in the whole oil fields. In other words, there is a gap between the water injection demand and the supply of water with respect to medium- and low-permeability reservoirs, but there is an excess of accumulated polymer-contained produced water in oil field ground systems. In addition, the typical characteristics of polymer-flooding produced water are high electronegativity, high viscosity, small particle size of oil droplets, high strength of water film, and strong emulsification tendency and stability [9–12]. Many problems arise during the process of treating polymer-flooding produced water, such as the purification effect, facility pollution, running stability, and economic rationality. Moreover, with the increase of polymer concentration, the complexity of the water quality characteristics makes these problems more prominent. In recent years, some innovative technologies for the treatment of polymer-flooding produced water, such as oxidation technology (chlorine dioxide), membrane filtration, biochemical treatment, biofilm, magnetic separation, air flotation, and suspended sludge have attracted attention and attempts.

However, due to investment, operation, and reconstruction, more practical situations are based on the existing two-stage process, making produced water filtration technology of oil field transition from stereotyping to serialization and individualization [13–16]. First, in reservoirs with different permeability, the targeted treatment of produced water with different water quality has been implemented, and attempts have been made to improve the utilization rate and operating load of the existing equipment. Second, in order to reduce the treatment load of filter tanks, gas flotation deoiling, swirl deoiling, and coarse granulation deoiling have been added to the settling section [17–25]. Third, when the purification effect of the filter tank is weakened, the intensity of backwashing is increased by experience, the scour of water flow on the surface of the filter medium is strengthened, and the intensity of rolling and friction of the filter medium is enhanced in order to improve the effect of backwashing and restore the filtration performance of the filter medium [26–28]. However, several problems, such as blockage, holding pressure, and large head loss, which are prone to occur during the operation of the filter tank and are caused by the change in water quality when produced water contains polymers, still need to be solved urgently [29,30].

The optimization of an oil–gas production system is important for system efficiency improvement, and the valuable layout optimization methodologies of large-scale oil–gas gathering system were proposed by Liu et al. [31]. In the optimization and design of produced water treatment system, the transportation and attachment of oil droplets and suspended solids to the filter medium constitutes the main process of filtration treatment, and it is also the core process for realizing the advanced treatment of produced water [32–37]. The current filtration process and operation parameters of oil fields are constructed and designed for water-drive produced water. In the treatment of polymer-flooding produced water, improper operation parameters will directly affect the water quality, the anti-pollution capabilities of the filtration equipment, and the operational stability of the system. In the final stage of produced water treatment, the effect of oil and suspended solids removal by the filter tank directly affects the quality of water injection [38]. Therefore, the primary research priorities are to study the composition of the pollutants intercepted by the filter medium from polymer-flooding produced water, to analyze the influence of the presence of polymers on oil and suspended solids removal, and to determine why the filtered water quality is not meeting standards. This is the basis of our research on the optimization of water quality, equipment, and operation parameters in the treatment of polymer-flooding produced water. This research will be useful to improve the operation of produced water treatment systems in oil fields that utilize polymer-flooding.

2. Materials and Methods

2.1. Design of the Filter Simulation Device

In order to deeply analyze the oil pollution characteristics of the filter layer and study the causes and mechanisms of filter medium pollution, it is necessary to analyze the structure, principle, and technology of the filter tank used in an oil field produced water station. According to this, we designed and built the filtration simulation device, and conducted a laboratory experiment on polymer-flooding produced water filtration.

In the oilfield produced water treatment process, the commonly used filter tank structure is shown in Figure 1a. The produced water separated by the gravity sedimentation of the upper stage flows into the radial water distribution pipe from the inlet, thereby performing uniform water distribution, and flow passes through the grading filter layer to achieve the purpose of removing oil droplets and removing suspended solids. The key to the effectiveness of the filtration process is the gradation form of the filter layer. Then the purified water exits the filter tank from the dendritic water collection pipe. The support layer near the water collection pipe is filled with large-sized pebbles. At the same time, the function of supporting the water collection pipe structure and the uniform water collection in porous medium space is realized; this creates a good flow field for the filtration process.

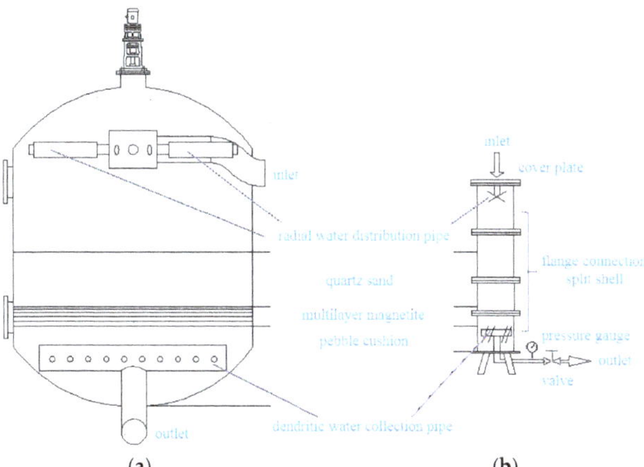

Figure 1. Schematic illustration of the filtration device for treating polymer-flooding produced water: (**a**) oil field filter tank; (**b**) filtration simulation device.

In order to simulate the filtration process of the polymer-flooding produced water realistically, and analyze interception characteristics of the filter layer and the pollution mechanism of the filter medium, it is necessary to consider that the filter tank has uniform water distribution, uniform water collection, and the gradation form of the filter layer is consistent. This ensures that the designed filtering simulation device and the filter tank in the oil field produced water station are geometrically similar and flow similar.

As shown in Figure 1, the filter tank in the oil field produced water treatment station has a diameter of 4000 mm and a total height of 4500 mm. For geometric similarity, the filter layer, which actually plays a filtration role, is mainly considered. The designed filtration simulation device diameter is 1/10 of the diameter of the oil field filter tank, and the height and grading form of the filter layer are consistent with the filter tank in the oil field. The filter medium ratio filling in the filter layer (in order from top to bottom) is shown in Table 1.

Table 1. Filter medium proportioning and gradation.

Number of Filter Layer	Filter Medium Type	Particle Diameter	Thickness
1	quartz sand	φ0.8 mm	800 mm
2	magnetite	φ0.25–0.5 mm	50 mm
3	magnetite	φ1–2 mm	50 mm
4	magnetite	φ2–4 mm	100 mm
5	magnetite	φ4–8 mm	100 mm
6	magnetite	φ8–16 mm	100 mm
7	pebble cushion	φ16–32 mm	300 mm (Filtration simulation device) 1300 mm (Oil field filter tank)

For flow similarity, analyze and consider according to Equation (1).

$$v = Q/A \tag{1}$$

where, v—filtering velocity;

Q—incoming water quantity;

A—effective filtration cross section area.

On the basis of filter medium proportioning and gradation consistency, by using the same filtering velocity as the flow similarity criterion, it can ensure that the flow state in the filtration simulation device and the oil field filter tank is the same.

In the design of the device, the problems of pressure bearing, uniform water distribution, uniform water collection, and boundary layer influence are considered respectively. The device is made of plexiglass material, the diameter of the radial water distribution pipe is 20 mm, and the diameter of the dendritic water collection pipe is 10 mm. The split shell is connected by a flange, the thickness is 10 mm, the total height is 2000 mm, the diameter is 400 mm, and the cover plate above the shell is made of phenolic resin.

Compared with the actual filter tank in an oil field, the diameter of the filtration simulation device was only 1/10 of the size, so there must have been a boundary layer with a fast flow velocity near the tank wall. In order to eliminate the influence of the boundary layer on the filtration simulation, the 5 mm area close to the inner wall was compacted when the filter medium was filled. On the inner wall of the device, a strip with a thickness of 3 mm was affixed to the inner wall every 200 mm of the device to block the flow of the boundary layer. In addition, the gasket at the flange joint protruded inward by about 3 mm, further blocking the flow of the boundary layer. The designed pressure resistance of the filter tank in the polymer-flooding produced water station was 0.8 MPa, the maximum pressure was 0.7 MPa in actual operation, and the common pressure was 0.65 MPa. The consistency of the filtration pressure difference was mainly considered in the establishment of the simulation device. Figure 2 shows the filtering simulation device.

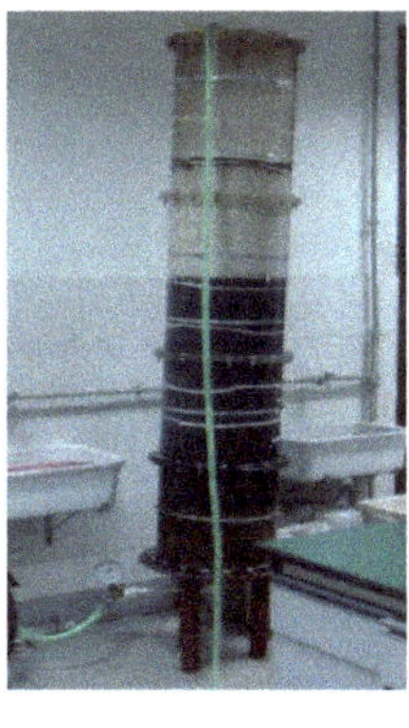

(a)

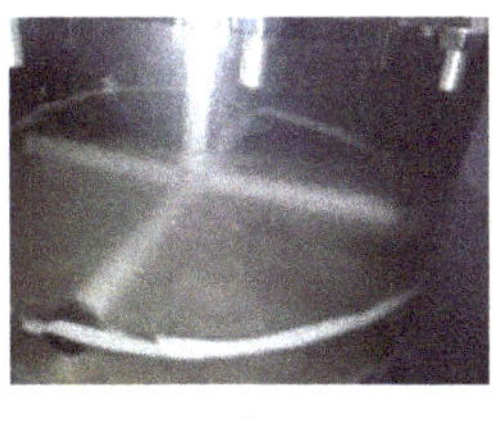

(b)

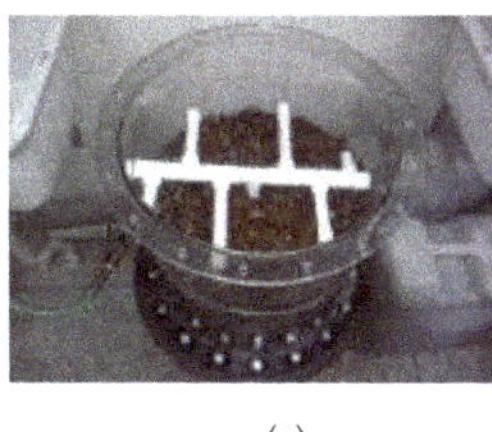

(c)

Figure 2. Filtration simulation device for treating polymer-flooding produced water: (**a**) whole shape; (**b**) top water distribution pipe; (**c**) bottom water collection pipe.

2.2. Experimental Water Quality

A water sample at the inlet of the filter tank of the oil field produced water station was used for the laboratory experiment. The raw water contained a polymer concentration of 514 mg/L, the oil content was 79.1 mg/L, and the suspended solids content was 53.7 mg/L. Among them, oil content and suspended solids content were analyzed by means of the spectrophotometry method and the gravimetric method, respectively [39,40], and polymer concentration was analyzed by means of the starch/cadmium iodide method [41]. The water sample was stored in the mixing tank to ensure the uniform dispersion of oil and water. The filtration period was 24 h, and the flow rate of the pump was adjusted to ensure that the Reynolds number of the laboratory device was similar to that of the field device.

3. Results and Discussion

3.1. Analysis of the Pollution Characteristics of the Filter Medium

As shown in Figure 3, after a filtration cycle, the appearance of the filter medium became black overall, and in particular, a great deal of black-brown oil pollution was intercepted on the surface of the filter layer.

The intercepted oil pollution was tested and analyzed by SY/T 5119-2016 "Analysis of soluble organic compounds and crude oil components in rocks", GB/T 8929-2006 "Determination of water content in crude oil by distillation", SY/T 0600-2016 "Oil field water scaling trend prediction", and other related standards. The results are shown in Table 2.

Figure 3. Intercepted oil pollution on the surface of the filter layer.

Table 2. Oil pollution components.

Sample Name	Water Content (%)		Waste Oil (%)		Solid Content (%)
Oil pollution intercepted by the filter medium	26.65		71.00		2.35
	saturated hydrocarbon (%)	arene (%)	asphaltene (%)	colloid (%)	non-hydrocarbon (%)
	51.28	20.39	15.93	11.83	12.4

From the analysis of the oil pollution composition, as shown in Table 2, it can be seen that waste oil was the main component of the oil pollution intercepted by the filter medium, accounting for 71%, whereas the solid phase content constituted only 2.35%. Components of colloid and asphaltene were present in the waste oil, accounting for nearly 30%. Thus, the carrying capacity of the polymer-flooding produced water with respect to waste oil was very strong and the difficulty of treating the produced water in this first stage settlement treatment was greatly increased.

Table 3 shows the solid phase composition in the oil pollution, combined with the microscopic morphology as shown in Figure 4, it can be seen that the intercepted oil pollution on the surface of the filter medium was a whole formed by the accumulation and cementitious formation of fine particles. The main elements in the solid phase inorganic composition were Si, O, C, Fe, Al, and Ca, containing 69.69% aluminosilicate (silica, alumina), 12.05% carbonate (calcium carbonate, magnesium carbonate, barium carbonate, etc.), and 9.09% iron oxide. The results show that the components intercepted by the filter layer were crude oil and silicon aluminate scale carried in the process of oil field production, as well as iron oxides formed by the corrosion of pipelines and equipment during transportation.

Table 3. The solid phase composition of the oil pollution.

Composition	Content (%)	Composition	Content (%)
SiO_2	54.0100	P_2O_5	0.2578
Al_2O_3	15.6774	Cr_2O_3	0.0316
Fe_2O_3	9.0906	MnO	0.1694
$BaCO_3$	5.7192	Co_2O_3	0.0272
$CaCO_3$	4.1563	NiO	0.0134
$MgCO_3$	2.1782	CuO	0.0198
Na_2O	1.2061	Ga_2O_3	0.002
K_2O	0.8665	Rb_2O	0.0025
$SrSO_4$	0.2587	WO_3	0.1524
TiO_2	0.2475	PbO	0.0132
ZnO	0.2394	Y_2O_3	0.0008

Figure 4. Micromorphology of the oil pollution on the filter medium surface (amplification factor: 3000).

3.2. The Micromorphology of the Polymers in the Filtration Process

3.2.1. Polymer Concentration Analysis

The polymer concentration in the unfiltered produced water, filtered water, and filter medium after one filtration cycle was analyzed using a starch cadmium iodide method. The results are shown in Table 4.

Table 4. Polymer concentration.

Unfiltered Produced Water (mg/L)	Filtered Water (mg/L)						Filter Layer (mg/L)
	4 h	8 h	12 h	16 h	20 h	Average Value	
514	22	96	223	402	453	238	907

The analysis shows that there was a significant difference between the polymer concentration in the unfiltered produced water and the filtered water, and the polymer concentration increased from 514 mg/L in the unfiltered produced water to 907 mg/L in the filter medium. Since polymers are not easily dissolved into the filter medium solid and are easily adsorbed on the surface of the filter medium, the polymer concentration adsorbed on the surface of the filter medium is much higher than that of the produced water, which indicates that the filter medium surface can adsorb some of the polymer concentration from the produced water. At the five moments of sampling, the polymer concentration after filtration was lower than that in the unfiltered produced water, which indicates that the filter medium had an adsorption effect on the polymers. However, it can also be observed that the polymer concentration in the water samples after 4 h of filtration was only 22 mg/L, indicating that the filter medium was very effective in the early stage of filtration and the adsorption capacity was the best at this time. With the gradual prolongation of filtration time, the polymers adsorbed on the surface of the filter medium reached saturation, so the polymer concentration in the filtered water increased gradually. As these polymer molecules began to flow out of the filter tank, the small particle size oil droplets and suspended solids also lost the opportunity to contact the surface of the filter medium and were carried out of the tank with the water flow.

3.2.2. Morphology and Distribution of Residual Polymers Before and After Filtration

The filtration simulation experiment was carried out by using the filtration simulation device, and sampling analysis was performed every 4 h. From the appearance characteristics of the water samples before and after filtration in Figure 5, it can be seen that the produced water before filtration was turbid, and in it, there were many suspended solids and emulsified oil droplets. After filtering for 4 h, the filter medium was relatively clean, because it was in the early stage of filtration; therefore, the interception and adsorption effect of the filter medium on the oil droplets and suspended solids was very high, and the filtered water was also very clear. As the filtration continued, the filter medium intercepted too many oil droplets and suspended solids, polluting the filter medium bed and the filtered effluent, and the adsorption and interception capacity were weakened. Therefore, the color of the water samples from 8 to 20 h after filtration gradually darkened in appearance. Thus, the filtration effect of the filter medium on the unfiltered produced water was obvious, but with the prolongation of filtration time, secondary pollution emerged, and the effectiveness of the treatment worsened.

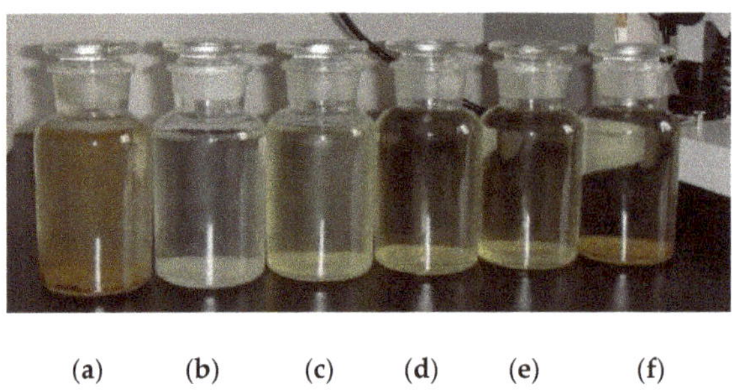

(a) (b) (c) (d) (e) (f)

Figure 5. Appearance of the polymer-flooding produced water before and after filtration. (**a**) unfiltered produced water; (**b**) 4 h; (**c**) 8 h; (**d**) 12 h; (**e**) 16 h; (**f**) 20 h.

Figure 6 shows the SEM image of the polymer-flooding produced water tested by freeze-drying before and after filtration. It can be seen that the morphology of the linear polymer molecules in the produced water before filtration was vague and difficult to distinguish and identify; the polymers linear molecules appeared as a blurred white streak. The appearance of this morphological characteristic is due to the high salinity of the produced water, which caused the dried polymers to adsorb a large amount of inorganic salt, and thus, the morphology of the polymer molecules was obscured by the inorganic salt. In the initial stage of filtration, after 4~12 h of filtration, most of the polymers in the produced water were adsorbed and intercepted by the filter medium bed; therefore, the water samples freeze-drying at the outlet after filtration were mainly composed of inorganic minerals, and the crystal structure became very clear. Moreover, polymer linear molecules could not be observed. It was revealed that the polymer molecules with large viscosity attached easily to the surface of the filter medium. However, with the prolongation of the filtration time, after 16–20 h of filtration, the adsorption of the polymers in the filter medium bed gradually reached saturation, and the polymer concentration in the filtered water gradually increased. The polymers adsorbed on the surface and filled in the mineral crystal, making the crystal morphology gradually appear blurry. At the same time, the linear network of polymer molecules gradually appeared.

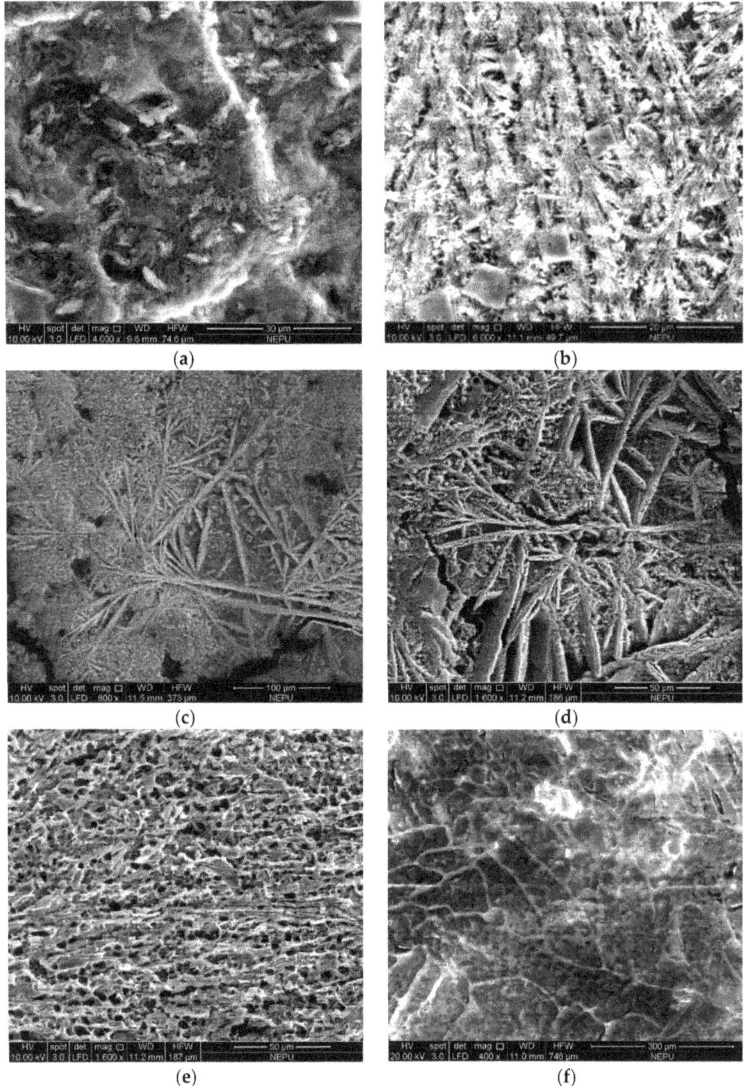

Figure 6. Micromorphology of polymer flooding produced water before and after filtration: (**a**) before filtration; (**b**) after filtration for 4 h; (**c**) after filtration for 8 h; (**d**) after filtration for 12 h; (**e**) after filtration for 16 h; (**f**) after filtration for 20 h. Amplification factor: 3000.

3.2.3. Microcosmic Characteristics of Intercepted Polymers in the Filter Layer

In the simulation experiment, when filtration was carried out for 23 h, the pressure difference increased to 0.007 MPa, and the effluent flow rate became smaller, indicating that the surface interception and internal absorption of the filter medium bed were saturated, which was regarded as the completion of a filtration cycle. At this point, the filtration simulation ended, the top cover of the filter tank was opened, the layered sampling of quartz sand filter medium (upper, middle, and lower) was carried out, and the surface micromorphology of the wet filter medium was analyzed.

As shown in Figure 7, it can be seen from the microscopic morphology of the polymers on the wet filter medium that the polymers were enriched due to their adsorption on the surface of the filter medium; therefore, the surface of the filter medium particles was coated with thick polymers. This promoted the decrease of the adsorption performance of the filter medium, affected the adsorption of the filter medium with respect to the emulsified oil droplets and suspended solids, and reduced the filtration effect and the filtration efficiency. Moreover, the adsorption of the polymers was a kind of dynamic adsorption. As the polymer mucous membrane thickened, the water phase continuously washed away the old membrane, and then, a new membrane was formed. When the water phase washed away the polymer mucous membrane, it also carried away some oil droplets and suspended solids attached to it, thus worsening the filtration effect.

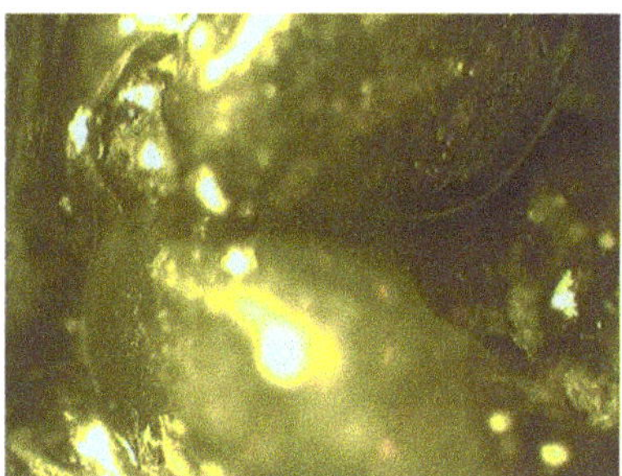

Figure 7. Metallograph of polymers adhered on the wet filter medium.

Similarly, the surface morphology of the polluted filter medium was analyzed by SEM to describe the fouling characteristics and pollution mechanism of the filter layer. Figure 8 shows the SEM images of the upper, middle, and lower quartz sand filter after filtration.

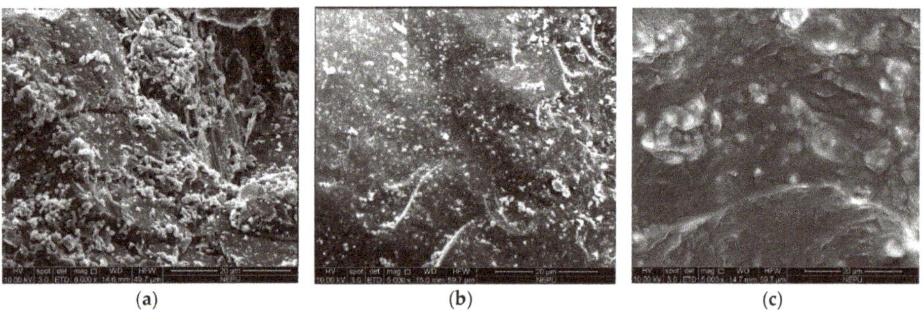

(**a**) (**b**) (**c**)

Figure 8. Micromorphology of quartz sand after filtration: (**a**) upper filter layer; (**b**) central filter layer; (**c**) bottom filter layer. Amplification factor: 6000.

It can be seen that in the upper layer, after the polymer-flooding produced water flowed through the surface of the filter medium, there were flocs on the surface of the angled quartz sand filter medium, and many types of solid particles were adsorbed. At a larger multiple, the adhesion of the suspended solids particles to the surface of the filter medium was more clearly visible, the composition of the suspended solids in the produced water was observed to be complex, and the shape of the suspended solids which adhered to the surface of filter medium was different. In the middle layer, the deposit on the surface of the filter medium was an inorganic/organic complex, and the particles were large, dense, and strong in adhesion. The results show that the long chain of polyacrylamide was broken due to friction and shear when the produced water passed through the filter medium, resulting in the decomposition of the polymers, adhesion to the rough inorganic deposit surface, and formation of an organic deposition layer together with the crude oil in the produced water. In the lower layer, there was a small amount of calcite crystal and flake aragonite on the surface of the filter medium, and no polymer deposition was seen. The results show that with the pore flow of the polymer macromolecules carried by the water phase between the filter medium, the angled quartz sand filter medium pulled off the long chain of polymer molecules, and the polymer molecules changed from long-chain macromolecules to relatively short-chain molecules, which are not easily adhered to the filter medium. Therefore, more of the smaller inorganic deposit crystals were deposited here, and the state was irregular. Under the influence of the flow, the inorganic deposits did not easily form large volume deposits. Although the long chain of polymer molecules became a short chain, it still maintained high polymer concentration, and the longer the filtration time, the higher the polymer concentration in the filtered water. The short-chain and high-concentration polymers still carried oil droplets and suspended solids in the filtered water, which affected the filtration interception and adhesion function.

Figure 9 is the computed tomography scanning diagram of the distribution state of the upper filter medium particles. The blue/blue-purple block area shows the filter medium particles, and the dark pink area shows the suspended solids particles. It can be further seen that there were more suspended solids on the surface of the filter medium, but there were also a lot of suspended solids in the middle of the pores, and no obvious "no suspended solids zone" was formed in the middle of the pores. This indicates that the adsorption of the suspended solids on the surface of the filter medium was a basis, and the suspended solids adhered to each other, connected with each other, and finally filled the pores, which was also the main reason for the blockage of the filter medium and the increasing pressure drop in practical operation.

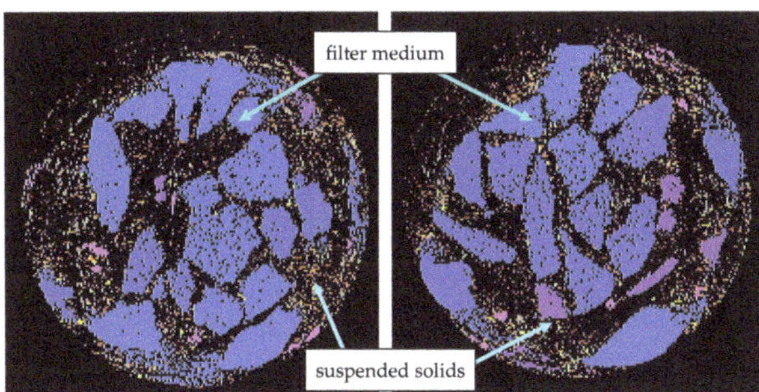

Figure 9. Computed tomography scanning of the oil pollution in the upper filter layer.

3.3. *Analysis on the Causes of Substandard Filtered Water Quality*

3.3.1. Analysis on the Causes of Substandard Suspended Solids in the Filtered Water

The failure of the filter medium to intercept and adsorb the suspended solids in the water is the fundamental reason why the filtered water does not meet permissible standards. The water flow carries suspended solids in the channel of the filter medium pores, and the size of the pores has a certain influence on the interception effect. From the point of view of the relative size of the pores and the suspended solids, the reasons for the high amount of suspended solids are analyzed below.

There are two kinds of maximum probability of filter medium particle stacking: triangular stacking and square stacking. In this study, the pore inscribed circle method was used to calculate the minimum diameter of the pores formed by these two stacking methods, as shown in Table 5.

Table 5. Filter medium porosity in contrast with the manufacturer's standard.

Petroleum Industry Standard SY/T 5329-2012 Allowable Median Particle Size		Minimum Pore Interception Capacity of the Filter Medium (According to the 1/3 Bridge Principle)	
maximum value	minimum value	triangular stacking	square stacking
4.0 μm	2.0 μm	12.9 μm	34.5 μm

(1) Triangular Stacking

As shown in Figure 10, according to the geometric relationship of the inscribed circles in the pores, the radius of the inscribed circle is $r = \frac{2}{3}\sqrt{3}R - R$; therefore, the pore diameter is $d = 2r = 0.3094\ R$. If the particle diameter is set to D, then $R = D/2$; therefore, the pore diameter is $d = 0.1547\ D$.

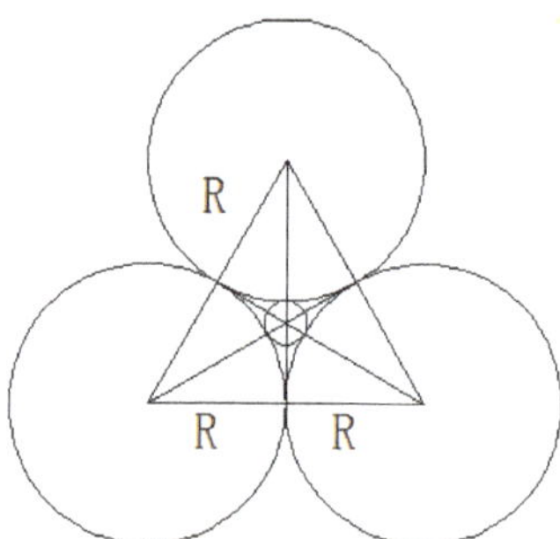

Figure 10. Triangle stacking mode.

(2) Square Stacking

As shown in Figure 11, an inscribed circle is made inside the pore, and the diameter of the inscribed circle is taken as the pore diameter. According to the geometric relationship, the diameter of the inscribed circle is as follows: $d = (\sqrt{2} - 1)\ D = 0.414\ D$.

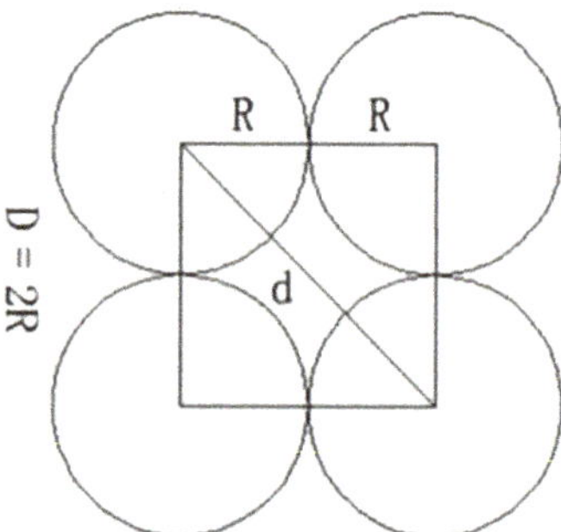

Figure 11. Square stacking mode.

The interception effect of the filter medium on the suspended solids can be divided into screening and adsorption. The screening effect is aimed at the larger suspended solids particles, which can be intercepted on the surface of the filter layer, because they cannot pass through the filter layer. Although the smaller suspended solids particles can enter the filter layer, these particles are adsorbed in the filter layer and filtered out, which is the adsorption effect.

Under this kind of filter medium gradation form, the particle size of the quartz sand filter medium was 0.8 mm, and the pore diameters of the triangular and square stacking modes were 123.76 and 331.2 μm, respectively. According to the 1/3 bridge principle, the minimum diameter of the suspended solids that could be intercepted by these two stacking methods was 41.25 and 110.4 μm, respectively.

There was a magnetite layer with a smaller particle size in the filter medium gradation. According to the minimum particle size of 0.25 mm, the pore diameters of the triangle and square stacking modes were 38.675 and 103.5 μm, respectively. According to the 1/3 bridge principle, the minimum diameter of suspended solids that could be intercepted by these two stacking methods was 12.9 and 34.5 μm, respectively.

As shown in Table 5, according to the standard SY/T 5329-2012 "Recommended index and analysis method for water injection quality of a clastic rock reservoir", the requirement for the median particle size of the suspended solids to maintain water injection quality is 2.0~4.0 μm. If only the bridge interception capacity is relied upon, small enough particles of the suspended solids cannot be removed, and the filtered water cannot meet the standard. Therefore, the effective removal of the suspended solids by the filter medium cannot rely only on the bridge interception capacity of the filter layer but must also rely on the adsorption capacity of the inner surface of the filter layer.

The particle size distribution of the suspended solids in the unfiltered produced water was determined by a laser particle size analyzer. In Figure 12, the results show that the particle size range of the suspended solids in the unfiltered produced water was 1.21~116.80 μm, and the median particle size was 19.93 μm. In Figure 13, it can be seen that the particle size range of the suspended solids in the filtered water was 0.43~62.63 μm, and the median particle size was 10.06 μm.

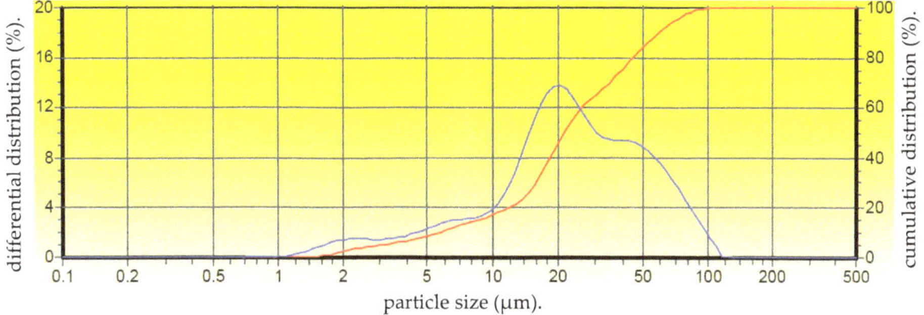

Figure 12. Particle size distribution of the suspended solids in the unfiltered produced water.

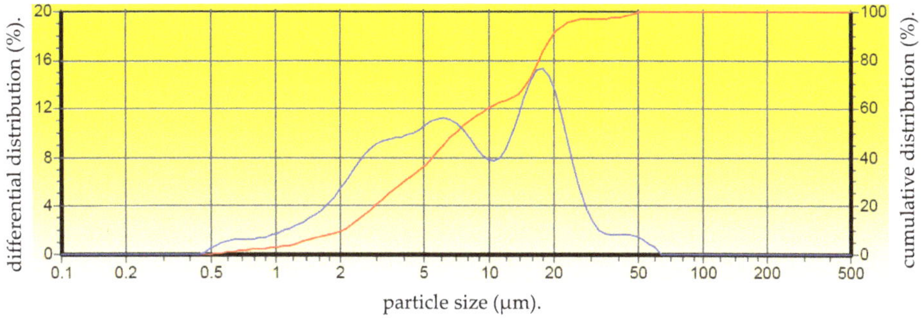

Figure 13. Particle size distribution of the suspended solids in the filtered water.

Between the two kinds of filter packing methods—triangle and square the pores formed by square stacking were larger (34.5 μm), indicating that the theoretical minimum suspended solids removal rate was 33.63%. However, the maximum particle size of the suspended solids in the filtered water was 50.88 μm, indicating that suspended solids particles larger than this particle size were filtered out and the actual removal rate was 14.72%. The filter medium not only removed the suspended solids by interception, but the small particles were also adsorbed by the filter medium in the filter layer. The removal rate should have been greater than 33.63% but was actually only 14.72%, indicating that there were many adverse effects on the removal of the suspended solids under the conditions of polymerization.

According to the data and analysis, the reasons that the content and particle size of the suspended solids in the filtered water exceeded the standard can be summarized as follows: (1) The porosity of the filter layer was large, only 20%~30% of particles of sufficient size can be intercepted by the bridge principle, and other particles that are smaller than the pores must rely on the adsorption of the filter medium surface in order to be intercepted. (2) There is competitive adsorption between the polymers and the suspended solids on the surface of the filter medium, and the affinity of the filter medium surface to the polymers is greater than that to the suspended solids, which limits the adsorption capacity on the surface of the filter medium with respect to the suspended solids.

3.3.2. Analysis on the Causes of Substandard Oil Content in the Filtered Water

1. Effect of the polymers

The relative positions of the oil droplets and the residual polymers between the filter medium were analyzed to investigate the accumulation and adhesion of the two in the filter medium. Figure 14 shows the adhesion characteristics of the polymers and oil droplets at different observation points of the filter layer. It can be seen that under ordinary light irradiation, the surface of the filter medium particles was bright, and the oil droplets or oil films were yellow-green under ultraviolet light irradiation.

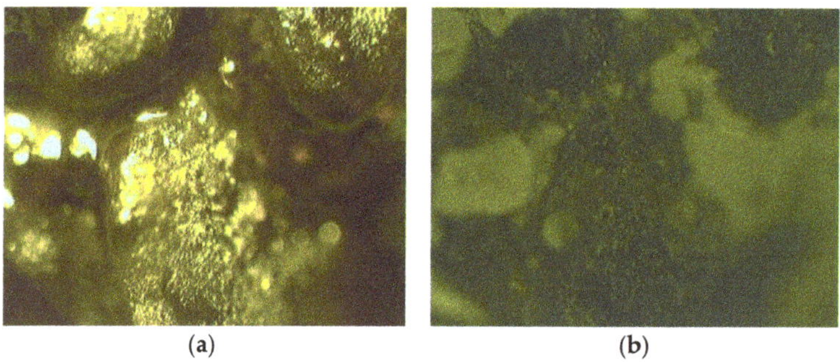

Figure 14. Micromorphological distribution of the polymer and oil in the filter layer: (**a**) polymers micromorphology; (**b**) crude oil micromorphology.

Obviously, the viscosity of the polymers was larger than that of the oil droplets, and the surface of the filter medium was mostly adhered to by the polymer molecules. Most of the crude oil in the filter layer was distributed in the pores of the filter medium particles in film or droplet form, which leads to the problem of water flow washing away the oil droplets or oil films in the pores. Even if part of the oil droplets adhere to the surface of the filter medium, because the viscosity of the polymer-flooding produced water is relatively larger and because of the replacement effect, the crude oil is easily replaced by the polymers on the surface of the filter medium and in the pores of the filter layer, resulting in the loss of opportunity for the oil droplets to sufficiently adhere to the surface of the filter medium.

2. Effects of the suspended particles of the biomass

The filter membrane was used to filter the polymer-flooding produced water to obtain the oil mud interceptor, which was further washed, and the crude oil, polymers, and inorganic mineralization ions were removed in order to obtain the suspended particles of the biomass.

As shown in Figure 15, according to the results of the scanning electron microscope analysis, there was a large amount of suspended solids with spatial reticular structure in the produced water, and the pores were very developed; they are supposed to be composed of dead bacteria, hyphae, and spores. Its chemical components were cellulose and lignin, which have strong adsorption capacity for crude oil, polymers, and mud. During the filtration process, these oils were carried into the filtered water by the suspended particles of the biomass, which is one of the reasons why the oil content of the filtered water exceeded the permissible standard.

Figure 15. Suspended solids after wash (amplification factor: 400).

4. Conclusions

(1) Due to the existence of the polymers in the produced water, a stable colloidal system was formed on the surface of the filter medium, which reduced the adsorption of the crude oil and suspended solids by the filter medium. The existence of the polymers also increased the viscosity of the water, promoted the emulsification of the oil pollution, and increased the difficulty of the filtration separation, thus affecting the filtration effect. As a result, the content of the oil and suspended solids in the filtered water was higher than the permissible standard.

(2) As the polymers in the produced water was adsorbed on the surface of the filter medium particles and spread into a thin film, the outline became blurred. In the initial stage of filtration, most of the polymers and oil pollution in the produced water were adsorbed and intercepted by the filter medium bed, and the filtered water mainly contained inorganic mineral ions. At the later stage of filtration, the adsorption of the polymers by the filter medium bed gradually reached saturation, and the polymers and oil pollution content in the filtered water increased gradually.

(3) The concentration and particle size of the suspended solids exceeded the permissible standard after filtration which is mainly due to the unreasonable size and filter medium gradation, and the competitive adsorption mechanism between the polymers and the suspended solids on the surface of the filter medium. However, the oil concentration of the filtered water exceeded the allowable standard, because the polymers have the replacement effect on the oil droplets in the pores and on the surface of the filter medium, and the biomass suspended particles composed of dead bacteria, hyphae, and spores had strong adsorption and carrying ability with respect to the oil droplets, which caused the suspended solids in the filtered water to exceed the standard while the oil droplets were simultaneously retained in the effluent.

(4) In order to improve the filtration efficiency and ensure that the filtered water quality meets the applicable standards, we plan to conduct future studies investigating ways to replace the aging filter medium in a timely manner and take measures to prevent mixing layer and running media between different layers of the filter medium, strengthen the cleaning of the filter medium regularly to ensure efficient filtration performance, and optimize the filter medium gradation to achieve a better overall filtration effect. Furthermore, some new oxidation technologies, such as the oxidation of polymer-flooding produced water with chlorine dioxide, thereby reducing the polymer concentration in produced water and re-modifying the water quality to the conventional treatment process, is a useful research direction of produced water treatment in polymer-flooding technology.

Author Contributions: Conceptualization, writing—original draft, investigation, X.W.; funding acquisition, supervision, X.X.; project administration, formal analysis, W.D.; methodology, Z.T.; writing—review and editing, C.H.; visualization, B.W.

Funding: This research was funded by the National Natural Science Foundation of China, grant number 51508573, 51674086.

Acknowledgments: This research was supported by the National Natural Science Foundation of China (Grant No. 51508573, 51674086).

Conflicts of Interest: The authors declare no conflicts of interest.

References

1. Li, J.X. *Polymer Flooding Ground Engineering Technology*; Petroleum Industry Press: Beijing, China, 2008; pp. 1–25.
2. Gao, B.; Jia, Y.; Zhang, Y.; Li, Q.; Yue, Q. Performance of dithiocarbamate-type flocculant in treating simulated polymer flooding produced water. *J. Environ. Sci.* **2011**, *23*, 37–43. [CrossRef]
3. Zhang, R.; Yu, S.; Shi, W.; Tian, J.; Zhang, Z. Optimization of membrane cleaning strategy for advanced treatment of polymer flooding produced water by nanofiltration. *RSC Adv.* **2016**, *6*, 28844–28853. [CrossRef]
4. Wang, Z.H.; Lin, X.Y.; Yu, T.Y.; Zhou, N.; Zhong, H.Y.; Zhu, J.J. Formation and rupture mechanisms of visco-elastic interfacial films in polymer-stabilized emulsions. *J. Dispers. Sci. Technol.* **2019**, *40*, 612–626. [CrossRef]
5. Nie, C.; Xu, L.; Gu, D.; Cao, G.; Yuan, R.; Wang, B. Toward efficient demulsification of produced water in oilfields: Solar step directional degradation of polymer on interfacial film of emulsions. *Energy Fuels* **2016**, *30*, 9686–9692. [CrossRef]
6. Zhang, J.; Jing, B.; Tan, G.; Zhai, L.; Fang, S.; Ma, Y. Comparison of performances of different type of clarifiers for the treatment of oily wastewater produced from polymer flooding. *Can. J. Chem. Eng.* **2015**, *93*, 1288–1294. [CrossRef]
7. Xia, Q.; Guo, H.; Ye, Y.; Yu, S.; Zhang, R. Study on the fouling mechanism and cleaning method in the treatment of polymer flooding produced water with ion exchange membranes. *RSC Adv.* **2018**, *8*, 29947–29957. [CrossRef]
8. Wang, Z.H.; Bai, Y.; Zhang, H.Q.; Liu, Y. Investigation on gelation nucleation kinetics of waxy crude oil emulsions by their thermal behavior. *J. Pet. Sci. Eng.* **2019**, *181*, 106230. [CrossRef]
9. Chen, H.X.; Tang, H.M.; Duan, M.; Liu, Y.G.; Liu, M.; Zhao, F. Oil-water separation property of polymer-contained wastewater from polymer-flooding oilfields in Bohai Bay, China. *Environ. Technol.* **2015**, *36*, 1373–1380. [CrossRef]
10. Ren, G.M.; Sun, D.Z.; Chunk, J.S. Advanced treatment of oil recovery wastewater from polymer flooding by UV/H_2O_2/O_3 and fine filtration. *J. Environ. Sci.* **2006**, *18*, 29–32.
11. Liu, A.; Liu, S.Y. Study on performance of three backwashing modes of filtration media for oilfield wastewater filter. *Desalin. Water Treat.* **2016**, *57*, 10498–10505. [CrossRef]
12. Ebrahimi, M.; Kovacs, Z.; Schneider, M.; Mund, P.; Bolduan, P.; Czermak, P. Multistage filtration process for efficient treatment of oil-field produced water using ceramic membranes. *Desalin. Water Treat.* **2012**, *42*, 17–23. [CrossRef]
13. Makhmudov, Z.M.; Saidullaev, U.Z.; Khuzhayorov, B.K. Mathematical model of deep-bed filtration of a two-component suspension through a porous medium. *Fluid Dyn.* **2017**, *52*, 299–308. [CrossRef]
14. Bai, B.; Wang, J.Q.; Zhai, Z.Q.; Xu, T. The penetration processes of red mud filtrate in a porous medium by seepage. *Transp. Porous Media* **2017**, *117*, 207–227. [CrossRef]
15. Sharafutdinov, R.F.; Bochkov, A.S.; Sharipov, A.M.; Sadretdinov, A.A. Filtration of live oil in the presence of phase transitions in a porous medium with inhomogeneous permeability. *J. Appl. Mech. Tech. Phys.* **2017**, *58*, 271–274. [CrossRef]
16. Nasre-Dine, A.; Ahmed, H.; Abdellah, A.; Wang, H.Q.; Gilbert, L.B.; Tariq, O. Porous media grain size distribution and hydrodynamic forces effects on transport and deposition of suspended particles. *J. Environ. Sci.* **2017**, *53*, 161–172.

17. Le, T.V.; Imai, T.; Higuchi, T.; Yamamoto, K.; Sekine, M.; Doi, R.; Vo, H.T.; Wei, J. Performance of tiny microbubbles enhanced with "normal cyclone bubbles" in separation of fine oil-in-water emulsions. *Chem. Eng. Sci.* **2013**, *94*, 1–6. [CrossRef]
18. Xu, H.X.; Liu, J.T.; Wang, Y.T.; Cheng, G.; Deng, X.W.; Li, X.B. Oil removing efficiency in oil–water separation flotation column. *Desalin. Water Treat.* **2015**, *53*, 2456–2463. [CrossRef]
19. Li, R.Y. Application of gas-assisted solvent flotation technique on oil-field polymer-bearing produced water. *Appl. Mech. Mater.* **2013**, *316–317*, 902–905. [CrossRef]
20. Chiavonefilho, O. Oil removal of oilfield-produced water by induced air flotation using nonionic surfactants. *Desalin. Water Treat.* **2015**, *56*, 1802–1808.
21. Bhaskar, K.U.; Murthy, Y.R.; Raju, M.R.; Tiwari, S.; Srivastava, J.K.; Ramakrishnan, N. CFD simulation and experimental validation studies on hydrocyclone. *Miner. Eng.* **2007**, *20*, 60–71. [CrossRef]
22. Bai, Z.S.; Wang, H.L.; Tu, S.T. Experimental study of flow patterns in deoiling hydrocyclone. *Miner. Eng.* **2009**, *22*, 319–323. [CrossRef]
23. Bergström, J.; Vomhoff, H. Experimental hydrocyclone flow field studies. *Sep. Purif. Technol.* **2007**, *53*, 8–20. [CrossRef]
24. Lim, E.W.C.; Chen, Y.R.; Wang, C.H.; Wu, R.M. Experimental and computational studies of multiphase hydrodynamics in a hydrocyclone separator system. *Chem. Eng. Sci.* **2010**, *65*, 6415–6424. [CrossRef]
25. Zhang, Y.; Gao, B.; Lu, L.; Yue, Q.; Wang, Q.; Jia, Y. Treatment of produced water from polymer flooding in oil production by the combined method of hydrolysis acidification-dynamic membrane bioreactor–coagulation process. *J. Pet. Sci. Eng.* **2010**, *74*, 14–19. [CrossRef]
26. Filtration & Separation Group. New filtration technology reduces backwash wastewater. *Filtr. Sep.* **2010**, *47*, 8.
27. Slavik, I.; Jehmlich, A.; Uhl, W. Impact of backwashing procedures on deep bed filtration productivity in drinking water treatment. *Water Res.* **2013**, *47*, 6348–6357. [CrossRef]
28. Henderson, A. Backwashing filtration systems. *Prod. Finish.* **2002**, *55*, 23.
29. Loderer, C.; Pawelka, D.; Vatier, W.; Hasal, P.; Fuchs, W. Dynamic filtration-Ultrasonic cleaning in a continuous operated filtration process under submerged conditions. *Sep. Purif. Technol.* **2013**, *119*, 72–81. [CrossRef]
30. Liu, Q.Y.; Dai, Y.; Luo, Y.; Chen, Y.L. Ultrasonic-intensified chemical cleaning of nano filtration membranes in oilfield sewage purification systems. *J. Eng. Fibers Fabr.* **2016**, *11*, 17–25. [CrossRef]
31. Liu, Y.; Chen, S.Q.; Guan, B.; Xu, P. Layout optimization of large-scale oil-gas gathering system based on combined optimization strategy. *Neurocomputing* **2019**, *332*, 159–183. [CrossRef]
32. Liu, G.; Zhang, F.; Qu, Y.; Liu, H.; Zhao, L.; Cui, M.; Ou, Y.; Geng, D. Application of PAC and flocculants for improving settling of solid particles in oilfield wastewater with high salinity and Ca^{2+}. *Water Sci. Technol.* **2017**, *76*, 1399–1408. [CrossRef] [PubMed]
33. Zhang, Z. The flocculation mechanism and treatment of oily wastewater by flocculation. *Water Sci. Technol.* **2017**, *76*, 2630–2637. [CrossRef] [PubMed]
34. Ottaviano, J.G.; Cai, J.X.; Murphy, R.S. Assessing the decontamination efficiency of a three-component flocculating system in the treatment of oilfield-produced water. *Water Res.* **2014**, *52*, 122–130. [CrossRef] [PubMed]
35. Guo, J.L.; Meng, J.; Li, G.P.; Luan, Z.K.; Tang, H.X. Physicochemical interaction and its influence on deep bed filtration process. *J. Environ. Sci.* **2004**, *16*, 297–301.
36. Wu, C.Y.; Wang, Y.N.; Zhou, Y.X.; Zhu, C. Pretreatment of petrochemical secondary effluent by micro-flocculation and dynasand filtration, performance and DOM removal characteristics. *Water Air Soil Pollut.* **2016**, *227*, 415. [CrossRef]
37. Si, S.; Yan, Z.; Gong, Z. Pilot study of oilfield wastewater treatment by micro-flocculation filtration process. *Water Sci. Technol.* **2018**, *77*, 101–107. [CrossRef]
38. Zamani, A.; Maini, B. Flow of dispersed particles through porous media-deep bed filtration. *J. Pet. Sci. Eng.* **2009**, *69*, 71–88. [CrossRef]
39. Weschenfelder, S.E.; Mello, A.C.C.; Borges, C.P.; Campos, J.C. Oilfield produced water treatment by ceramic membranes: Preliminary process cost estimation. *Desalination* **2015**, *360*, 81–86. [CrossRef]

40. Jordan, M.M.; Johnston, C.J.; Robb, M. Evaluation methods for suspended solids and produced water as an aid in determining effectiveness of scale control both downhole and topside. *SPE Prod. Oper.* **2006**, *21*, 7–18. [CrossRef]
41. Zhong, H.; Yang, T.; Yin, H.; Lu, J.; Zhang, K.; Fu, C. Role of alkali type in chemical loss and ASP-flooding enhanced oil recovery in sandstone formations. *SPE Reserv. Eval. Eng.* **2019**. [CrossRef]

Review

NMR Determination of Free Fatty Acids in Vegetable Oils

Maria Enrica Di Pietro [1,*], Alberto Mannu [1] and Andrea Mele [1,2]

[1] Department of Chemistry, Materials and Chemical Engineering "Giulio Natta", Politecnico di Milano, Piazza Leonardo da Vinci 32, 20133 Milan, Italy; alberto.mannu@polimi.it (A.M.); andrea.mele@polimi.it (A.M.)

[2] CNR-SCITEC Istituto di Scienze e Tecnologie Chimiche, Via Alfonso Corti 12, 20133 Milano, Italy

* Correspondence: mariaenrica.dipietro@polimi.it; Tel.: +39-02-2399-3045

Received: 9 March 2020; Accepted: 25 March 2020; Published: 31 March 2020

Abstract: The identification and quantification of free fatty acids (FFA) in edible and non-edible vegetable oils, including waste cooking oils, is a crucial index to assess their quality and drives their use in different application fields. NMR spectroscopy represents an alternative tool to conventional methods for the determination of FFA content, providing us with interesting advantages. Here the approaches reported in the literature based on ^{1}H, ^{13}C and ^{31}P NMR are illustrated and compared, highlighting the pros and cons of the suggested strategies.

Keywords: nuclear magnetic resonance; free fatty acids; acid value; NMR quantification; waste cooking oils; waste oil characterization

1. Introduction

Free fatty acids (FFA) are hydrolysis products of triglycerides (TG) in vegetable oils. In edible oils, their formation occurs primarily during the production and the storage procedures of the oil and, in general, during handling of the raw material. Deterioration processes in lipids are additional sources of FFA. For instance, short-chain FFA can arise from the secondary oxidation of unsaturated aldehydes as well as from the cleavage of lipid hydroperoxides [1,2]. The FFA concentration in vegetable oils depends on multiple factors, namely the quality and variety of raw material, collecting conditions, processing, storage, the age of the oil and deterioration status [3,4]. The amount of FFA is even higher in waste cooking oils (WCO), since high temperature and exposure to air occurring during frying promote the hydrolysis and oxidation of triglycerides and increase the content of FFA in the oil [5–8].

An established method to measure the content of FFA is the so-called "acid value" (AV), referred to as the amount of potassium hydroxide (milligrams of KOH) required to neutralize the acidic fraction in one gram of sample (mg(KOH)/g(oil)). Practically speaking, the AV is a volumetric method for the determination of FFA in lipids (fatty oils and waxes), that consists in the titration of the sample with a standardized ethanolic solution of KOH using phenolphthalein as an endpoint indicator [9].

The AV is an important parameter for quality assessment in the determination of FFA in lipids. Indeed, the content of FFA in vegetable oils affects their properties and has implications in many application fields. In edible oils, FFA are undesirable, as high FFA content results in increased losses during refining, poor flavour quality and stability of the finished edible oil and rancidity of the oil [10–12]. At an industrial level, FFA are usually removed from crude oil through a refining procedure, but the process is not 100% efficient. Hence, the quantitative determination of the AV plays a central role as a quality index in production, trading storage, and marketing of edible oils, and in official food and pharmaceutical control [1,3,11]. Both the European Pharmacopoeia and the Codex Alimentarius establish maximum values for AV in pharmaceutical products and edible oils.

The FFA content is also important for edible and non-edible vegetable oils as well as WCO, with applications in biodiesel production [5,8]. Biodiesel is composed of methyl/ethyl esters containing long chain fatty acids and can be obtained directly by a transesterification reaction of raw materials having a low content of FFA. In the case of oils with a high content of FFA, the transesterification reaction can give a saponification reaction, causing deactivation of the catalytic species and a decrease in biodiesel yield [8]. It has been reported that, for FFA levels higher than 5%, the separation between the biodiesel and glycerol is hindered by the presence of soap, with a consequent decrease in the yield of the final product [5]. Thus, the FFA content of the feed stocks is a crucial quality parameter. A maximum acid value of 0.5 mg(KOH)/g is fixed by European norms for biodiesel.

The identification and quantification of FFA is also a main target in the development of new sustainable processes for WCO recycling and management. These used vegetable oils can be transformed into valuable goods and employed as bio-lubricants, animal feed, additives for bio-asphalts and bio-concrete, or in the field of energy production both by direct burning and for biofuel synthesis. Regardless of the specific application, a pre-treatment step is needed, which changes the composition and, consequently, also the FFA content [13–15]. The determination of the composition of waste oils, including the analytical measurement of acidity, is then crucial to assess its quality [16] and thus to determine how to better exploit the raw material.

Unfortunately, the classical AV method has a number of drawbacks. It usually depends on a visual endpoint, so that its accuracy can be compromised, for instance, in the case of a strongly coloured sample. Although the introduction of technologies such as potentiometric endpoint determination significantly improved the method, it still suffers from several disadvantages: it is usually slow, requires a large sample size and large amounts of organic solvents and harmful chemicals, and lacks specificity [2,12].

Alternative techniques have been proposed [17], which give good correlations with the classical AV, such as pH-metry [18], gas chromatography [19–22], HPLC [11,23], FTIR [10,24,25], near IR [26,27], colorimetry [28–30], or flow injection analysis (FIA) [31–33].

Moreover, NMR spectroscopy (^{1}H, ^{13}C, ^{31}P) has been proposed as an alternative method for FFA determination in the analysis of oils and fats [2,34]. As strong points of the technique, one can mention that NMR spectroscopy is a non-destructive, non-invasive methodology; it is usually quick and straightforward, and yields information on the composition of a mixture in a single spectrum, without the need for derivatization or pre-treatment of the sample. Moreover, it usually requires negligible sample preparation and small quantities of organic solvents or reagents. Given that the experimental setup complies with specific requirements, NMR analysis is quantitative, meaning that the integral of a signal is directly proportional to the number of the corresponding nuclei. This allows us to directly obtain the molar concentration from the NMR spectrum. Moreover, the NMR approach is also useful for samples that cannot be analysed by chromatographic techniques due to heat sensitivity or other issues. Over the years, NMR has stood out as an efficient technique in fatty acid characterization, to derive the composition of fatty acids (the so-called fatty acid profile) and the ratio between different acyl groups, to classify the quality of edible oils, to study and monitor the oxidation and deterioration process, to assess the purity and authenticity or detect adulteration. Many reviews and book chapters are available in the literature on this wide topic [2,34–38]. Particularly important in fatty acid characterization is the distinction between saturated and unsaturated acyl moieties, as well as the determination of the degree of unsaturation (through parameters such as the iodine value (IV), an empirical number related to the total number of unsaturations). The usefulness of NMR spectroscopy in this respect has also been widely demonstrated [39,40].

In this minireview, the focus will be the application of NMR methodology to the determination of the content of FFA in vegetable oils. The reviewed methods are applicable to edible and non-edible oils, as well as WCO, where the determination of the FFA content requires more and more attention.

2. Discussion

2.1. ^{1}H NMR

Thanks to its favourable properties (spin number I = 1/2, high gyromagnetic ratio equal to 26.75×10^7 rad T^{-1} s^{-1} and natural abundance of 99.99%), the ^{1}H nucleus is the most sensitive NMR probe and is suitable for the identification and quantification of even minor components within a short experimental time [2]. ^{1}H NMR spectroscopy has been widely used to obtain insights into lipid classes, fatty acid composition, levels of unsaturation, and several minor compounds [2,41,42]. Vegetable oils are mainly made up of mixtures of triglycerides, with different substitution patterns according to the length, degree and type of unsaturation of the acyl chains, and by minor components including mono- and di-glycerides, fatty acids, sterols, vitamins, and others [43]. The ^{1}H chemical shifts of the triglycerides are well known and minor oil components can be detected by ^{1}H NMR when their signals are not overlapped with those of the main components. Unfortunately, because of the chemical similarity between the free fatty acids and the glyceride esters in the oil, the ^{1}H NMR signals resonate close together and build clusters [42]. Hence, the ^{1}H NMR spectrum of vegetable oils have a small number of characteristic peaks so that the quantification of the individual components in mixtures can be challenging. To get an idea, the ^{1}H chemical shifts of the main FFA found in WCO (Figure 1) are given in Table 1 [8,16,44].

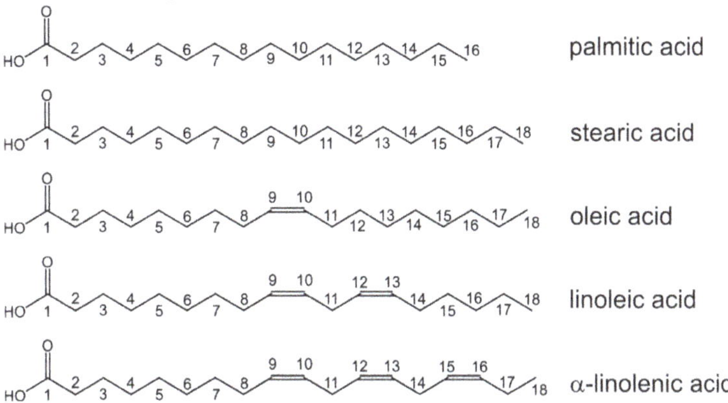

Figure 1. Structures of selected fatty acids present in waste cooking oils (WCO).

Table 1. ^{1}H chemical shifts of selected fatty acids present in WCO [44,45].

Proton Signal	Palmitic Acid (16:0)	Stearic Acid (18:0)	Oleic Acid (18:1 Δ^{9})	Linoleic Acid (18:2 $\Delta^{9,12}$)	α-Linolenic Acid (18:3 $\Delta^{9,12,15}$)
=CH-			5.36 (H9, H10)	5.37 (H9, H10, H12, H13)	5.36 (H9, H10, H12, H13, H15, H16)
=CH-CH_2-CH=				2.78 (H11)	2.80 (H11, H14)
-CH_2COOH	2.36 (H2)	2.35 (H2)	2.36 (H2)	2.36 (H2)	2.35 (H2)
=CH-CH_2-			2.03 (H8, H11)	2.06 (H8, H14)	2.04 (H8, H17)
-$CH_2$$CH_2$COOH	1.64 (H3)	1.63 (H3)	1.64 (H3)	1.64 (H3)	1.61 (H3)
-CH_2-	1.24 (H4–H15)	1.25 (H4–H17)	1.30 (H4–H7, H12–H17)	1.35 (H4–H7, H15–H17)	1.31 (H4–H7)
-CH_3	0.89 (H16)	0.88 (H18)	0.89 (H18)	0.90 (H18)	0.98 (H18)

Satyarthi et al. developed a ^{1}H NMR method to quantitatively determine FFA in vegetable oils, animal fats and biodiesel [46]. It is based on the integration of the signal corresponding to the α-carbonyl methylene protons of FFA (the methylene group directly adjacent to the COOH group) and the α -carbonyl-CH_2 signal of esterified fatty acids. In vegetable oil and biodiesel, α-CH_2 peaks of fatty acids appear at chemical shift (δ) values higher than those of the ester, as a consequence of the stronger deshielding effect of the carboxylic group with respect to the ester group. Hence, one of the peaks of the triplet of FFA is visible outside the α-CH_2 region of the ester, while the other two peaks overlap with those due to the ester (Figure 2). This means that a sample of vegetable oil or biodiesel containing both FFA and ester shows a pseudo-quartet signal in the α-CH_2 region of the proton NMR spectrum and that the intensity of the peaks depends on the content of FFA in esters. The FFA content can be calculated from the unmerged peak of the FFA triplet using the following equation [46]:

$$\%\ \text{FFA} = \frac{4 \cdot I_{\text{unmerged}-\alpha-\text{CH}_2-\text{FFA}}}{I_{\text{total}-\alpha-\text{CH}_2-\text{FFA}+\text{ester}}} \cdot 100 \quad (1)$$

where $I_{\text{unmerged}-\alpha-\text{CH}_2-\text{FFA}}$ is the area of the unmerged FFA peak and $I_{\text{total}-\alpha-\text{CH}_2-\text{FFA}+\text{ester}}$ the total area of the α-CH_2 of both FFA and ester. The pre-factor 4 accounts for the fact that the triplet of the α-CH_2 group of the FFA has an intensity ratio of 1:2:1, so that the total area is four times the area of the single unmerged FFA peak.

Another option is to deconvolute the peaks of the ester and the acid, so that the FFA content can be determined according to the following equation [46]:

$$\%\ \text{FFA} = \frac{I_{\alpha-\text{CH}_2-\text{FFA}}}{I_{\text{total}-\alpha-\text{CH}_2-\text{FFA}+\text{ester}}} \cdot 100 \quad (2)$$

with $I_{\alpha-\text{CH}_2-\text{FFA}}$ being the area of the triplet corresponding to the α-CH_2 of FFA.

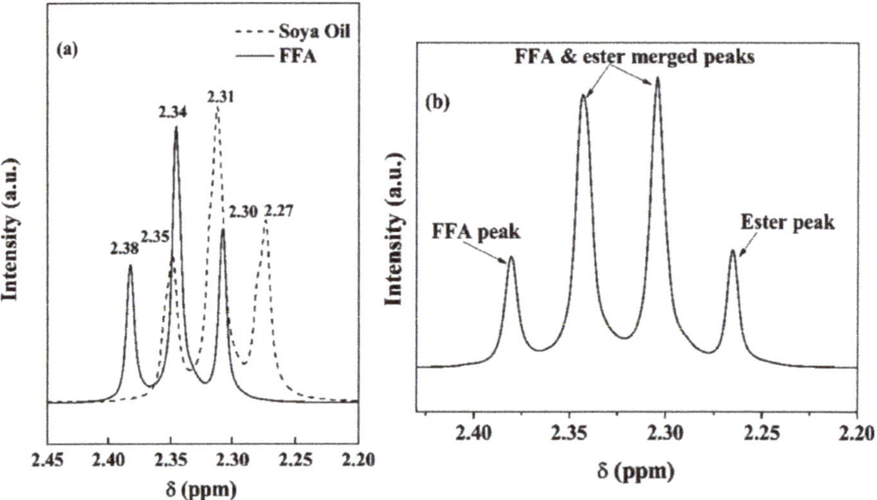

Figure 2. ^{1}H NMR spectrum of α-CH_2 region of (**a**) soybean oil and the free fatty acid (FFA) oleic acid, and (**b**) a mixture of oleic acid and its methyl ester. The sample for ^{1}H NMR was prepared by dissolving 20–25 mg in 0.6 mL of $CDCl_3$. The spectrum was acquired on a Bruker AV 200 MHz spectrometer at 298 K, using 30 scans, with an acquisition time of 3.9 s, a 30° flip angle and a relaxation delay of 1 s. Reprinted with permission from *Energy & Fuels*, Vol. 23, J. K. Satyarthi, D. Srinivas, P. Ratnasamy, "Estimation of Free Fatty Acid Content in Oils, Fats, and Biodiesel by ^{1}H NMR Spectroscopy" (pages 2273–2277). Copyright (2009) American Chemical Society.

The method was validated against conventional titration by measuring the FFA content in standard solutions of soybean oil and biodiesel added with known amounts of oleic acid. A good correlation of the calibration curves was observed, even though an intrinsic error of approximately 1% was reported [46]. To further test the performance of the ^{1}H NMR method versus titrimetric methods, the decrease in the oleic acid content was followed and quantified during its esterification reaction with methanol. Again, a good correlation in terms of oleic acid content at different intervals was observed [46]. A peculiar advantage of the proposed ^{1}H NMR method over the titrimetric method emerged in the quantification of FFA content in non-edible oils containing acidic impurities other than FFA. For instance, the ^{1}H NMR technique detected FFA content in karanja oil equal to 4.5 wt%, while a higher value (5.3 wt%) was estimated by titration. After transesterification into biodiesel, no FFA resulted by ^{1}H NMR analysis, whereas a residual acidity (0.79 wt%) was detected by the titrimetric method. These outcomes indicate that ^{1}H NMR spectroscopy is even more accurate than titration, since it detects exclusively the FFA content instead of the total acid value given by FFA and other acidic entities.

The main limitation of the approach by Satyarthi et al. is that it is designed for non-edible lipids and biodiesel with significant FFA amounts and may not be sensitive enough to detect small FFA contents present in commercially available edible oils or in pharmaceutical products.

Skiera et al. described an alternative rapid and simple ^{1}H NMR method that overcomes this limitation, which exploits the integration of the carboxyl group signal of FFA located in the downfield region of the spectrum [3]. The lipid sample is dissolved in a $CDCl_3$/DMSO-d_6 mixture (5:1 *v/v*) with tetramethylsilane as an internal reference and then dried over a molecular sieve. This sample preparation is necessary to reduce the fast proton exchange processes that occur in pure $CDCl_3$, which would cause line broadening or even the loss of the COOH signal. The quantification of FFA is based on the integration of the signal corresponding to the carboxyl proton at 11–12 ppm and the α-carbonyl CH_2 protons at 2.2–2.4 ppm (Figure 3). By normalising the α-carbonyl CH_2 signal to 6000, the number of those protons per TG unit, one obtains the FFA content in mmol/mol TG.

The efficacy and accuracy of the proposed ^{1}H NMR method was validated against AV on a wide range of oil varieties [1,3]. First, analysing by both techniques some standard solutions of rapeseed oil with known amount of palmitic acid, the authors derived a mathematical equation to convert the molar FFA amount determined by NMR into a parameter AV_{NMR} that is compared to the classical AV [3]. Then, more than 400 samples of different oil varieties (around 20) were analysed by both methods in two different studies and the results were similar for the majority of the samples [1,3]. Systematic deviations were observed for the pumpkin seed oils, with the classical AV method giving slightly higher values [3]. This is likely due to the strong colour of this oil, which complicates the visual endpoint determination in the AV method. The suitability for coloured oils represents, then, a strength of the ^{1}H NMR method compared to the classical AV method. Additional advantages of the ^{1}H NMR method are that it offers the opportunity to evaluate authenticity parameters out of the same ^{1}H NMR spectrum, such as the amount of hydroperoxides and aldehydes [3].

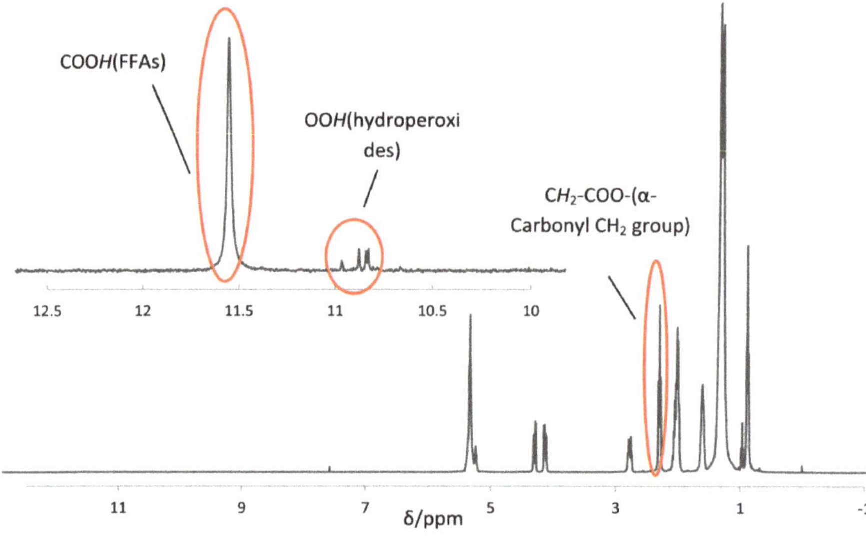

Figure 3. ^{1}H NMR spectrum of a rapeseed oil dissolved in a mixture of $CDCl_3$ and DMSO-d_6 (5:1, *v*/*v*), with enlargement of the carboxylic region. The spectrum was acquired on a Bruker AVANCE 400 MHz spectrometer at 300 K, using 128 scans, with an acquisition time of 7.97 s, a 30° flip angle and a relaxation delay of 1 s. Reprinted from the *Journal of Pharmaceutical and Biomedical Analysis*, Vol 93, C. Skiera, P. Steliopoulos, T. Kuballa, B. Diehl, U. Holzgrabe, "Determination of free fatty acids in pharmaceutical lipids by ^{1}H NMR and comparison with the classical acid value" (pages 43–50), copyright (2014), with permission from Elsevier.

However, the ^{1}H NMR method proposed by Skiera et al. presented some limitations in two cases. For hard fat samples, the AV_{NMR} values were significantly smaller than the AV ones [1]. This discrepancy was ascribed to the smaller average molecular weight of hard fat with respect to that of rapeseed oil, which was used for the development of the model equation. An adjustment of the equation was then made, including the integral of the COOH signal of an internal reference (1,2,4,5-tetrachloro-3-nitrobenzene, TCNB), leading to a good agreement with the classical AV method. Issues emerged also in the case of castor oil, where no COOH signal was detected in the ^{1}H NMR spectrum [1]. The effect here was due to the high content of a fatty acid with a hydroxyl functional group, the ricinoleic acid, whose easily exchangeable OH protons affected the fast proton exchange of the carboxylic protons. As a result, the amount of DMSO-d_6 used in the original sample preparation procedure was not enough to slow down proton exchange. To solve the issue, the sample preparation protocol for castor oil was modified by using a mixture of $CDCl_3$/DMSO-d_6 with a molar ratio of 2:1 *v*/*v*, obtaining a good correlation with the standard AV method.

Despite the potential spectral overlap, the ^{1}H NMR method found several applications. It was exploited, for instance, to estimate the FFA content in non-edible oils extracted from thirteen seeds from India [47], to follow the change in the amount of FFA and other components in different developmental stages of a non-edible oilseed, *Jatropha curcas* L. [48], or to detect and quantify all the different constituents, including FFA, coming from the hydrolysis of triglycerides during lipolysis [49]. More recently ^{1}H NMR was also applied to assess the quality especially in terms of the FFA content of biodiesel produced by transesterification of oil from *J. curcas* L. seeds [50].

To maximize the separation of the signals, very recently, San Martín et al. carried out a systematic investigation of different solvent mixtures to be used for the ^{1}H NMR analysis of edible fats and oils [51]. Among the mixtures tested, the best results were obtained using CCl_4/DMSO-d_6 or CS_2/DMSO-d_6/$CHCl_3$ as solvents. In these samples, an excellent separation of the signals of minor compounds in several edible oils (olive oil, sunflower oil, corn oil, sesame oil, peanut oil) and fats (butter, walnuts, salmon, dry sausage) was possible. An accurate detection and quantification of FFA was achieved with a good correlation between the FFA content estimated by NMR and that calculated by standard titration. As an additional benefit, these solvent mixtures allow the separation of signals with a minimum amount of deuterated solvents, which also means that they can be used in cases of lock and shim automation.

2.2. ^{13}C NMR

The main drawback of ^{1}H NMR spectroscopy in oil analysis is related to the small spectral width covered by protons. ^{13}C NMR spectra as an alternative to ^{1}H NMR offers some advantages [36]. First, ^{13}C NMR provides a large number of signals covering a wide range of chemical shifts, which makes the ^{13}C spectrum very informative. As an example, the ^{13}C chemical shifts of the main FFA found in WCO (Figure 1) are given in Table 2 [44]. Moreover, the low gyromagnetic ratio of the ^{13}C nucleus (6.728×10^{7} rad T^{-1} s^{-1}), its low natural abundance (1.07%) and low receptivity compared to ^{1}H (1.7×10^{-4}), are balanced by the low possibility of ^{13}C-^{13}C scalar coupling and by broadband heteronuclear decoupling. This results in sharp singlets for all ^{13}C absorptions, making it possible to measure small chemical shift differences [2]. Using the inverse gated decoupling technique to remove the nuclear Overhauser proton–carbon enhancement, and long pulse delays to ensure the complete relaxation of ^{13}C nuclei with long spin-lattice relaxation time, it is also possible to acquire a ^{13}C NMR spectrum under quantitative conditions, as in the case of ^{1}H NMR [36]. The main drawback is that ^{13}C NMR experiments typically need long measurement times to obtain a spectrum with proper signal-to-noise ratio. In order to shorten the total experimental time, it is possible to add relaxation reagents (typically chromium (III) acetylacetonate [Cr(acac)$_3$]), which are paramagnetic substances able to significantly reduce the relaxation times of ^{13}C nuclei without inducing relevant shifts.

Table 2. ^{13}C chemical shifts of fatty acids mainly present in WCO [44,45].

Carbon Signal	Palmitic Acid (16:0)	Stearic Acid (18:0)	Oleic Acid (18:1 Δ^{9})	Linoleic Acid (18:2 $\Delta^{9,12}$)	α-Linolenic Acid (18:3 $\Delta^{9,12,15}$)
-COOH	182.62 (C1)	182.78 (C1)	180.50 (C1)	180.16 (C1)	180.10 (C1)
=CH-			130 (C9, C10)	128-130 (C9, C10, C12, C13)	127-131 (C9, C10, C12, C13, C15, C16)
-CH_2COOH	36.68 (C2)	36.70 (C2)	33.96 (C2)	34.01 (C2)	33.95 (C2)
-CH_2-	32-34 (C4-C14)	29-32 (C4-C16)	29-31 (C4-C7, C12-C16)	29-31 (C4-C7, C15-C16)	29 (C4-C7)
=CH-CH_2-			27.12 (C8, C11)	27.25 (C8, C14)	27.22 (C8)
=CH-CH_2-CH=				25.65 (C11)	25-26 (C11, C14)
-CH_2CH_2COOH	27.33 (C3)	27.33 (C3)	24.59 (C3)	24.70 (C3)	24.56 (C3)
-CH_2CH_3	25.36 (C15)	25.36 (C17)	22.52 (C17)	22.54 (C17)	20.58 (C17)
-CH_3	16.78 (C16)	16.79 (C18)	14.07 (C18)	14.06 (C18)	14.27 (C18)

In the field of lipid chemistry and technology, high resolution ^{13}C NMR has established itself as a versatile technique to determine the composition of mixtures of fatty acids and lipid molecules, control the purity and reveal adulteration, authenticate the geographical origin and assess quality, quantitatively determine fatty acid profiles, evaluate oxidation products, analyse the phenolic fraction and determinate iodine value [2,37,38,52].

^{13}C NMR has also been applied in the determination of the free acidity in vegetable oils—that is, their FFA content. Indeed, the structure of the carbonyl region shows distinct signals for different mono-, di- and triglycerides and FFA in the mixture [35]. As shown in Figure 4, for a sample of virgin olive oil, the free carbonyls resonate at 176–178 ppm, while the esterified carbonyls resonate in the region 171–174 ppm [52]. By comparing the signal intensity of the two regions, the free acidity (% mole fraction) can be easily calculated by ^{13}C NMR as follows:

$$\%\ \text{FFA} = \frac{I_{\text{FFA}}}{I_{\text{FFA}} + I_{\text{EC}}} \cdot 100 \quad (3)$$

where I_{FFA} is the integral of the FFA region and I_{EC} is the integral of the esterified carbonyl region.

However, attention must be paid in order to obtain an accurate quantification, since carboxyl and carbonyl carbon atoms have significantly longer relaxation times than protons. Note, for instance, that T_1 relaxation times of 10–20 s were calculated for the carbonyl carbons of acyl groups of the different components present in palm oil (in $CDCl_3$ at 301 K) [53]. To obtain quantitative data, the relaxation issue can be bypassed by adding a relaxation reagent, e.g., [Cr(acac)$_3$], or by running ^{13}C NMR spectra with sufficiently long relaxation delay [2]. Although a validation test of the quantitative measurements run on virgin olive oil is not provided [52], the accuracy of the method was previously tested on samples of fish oils, which show a similar pattern in the carbonyl region [54]. The total FFA content of seven samples of tuna lipids was determined using ^{13}C NMR and compared to the values obtained using a UV method based on the formation of a complex with (AcO)$_2$Cu-pyridine. The good linearity observed between the two methods was then assumed to be a validation of the ^{13}C NMR approach, which was applied also to vegetable oils.

The strong discriminating power of the ^{13}C NMR method in the carbonyl region was demonstrated, for instance, in the estimation of the amount of FFA in non-edible oils extracted from various seeds from India [47], or in oil extracted from commercial samples of grated mullet roe (bottarga) [55].

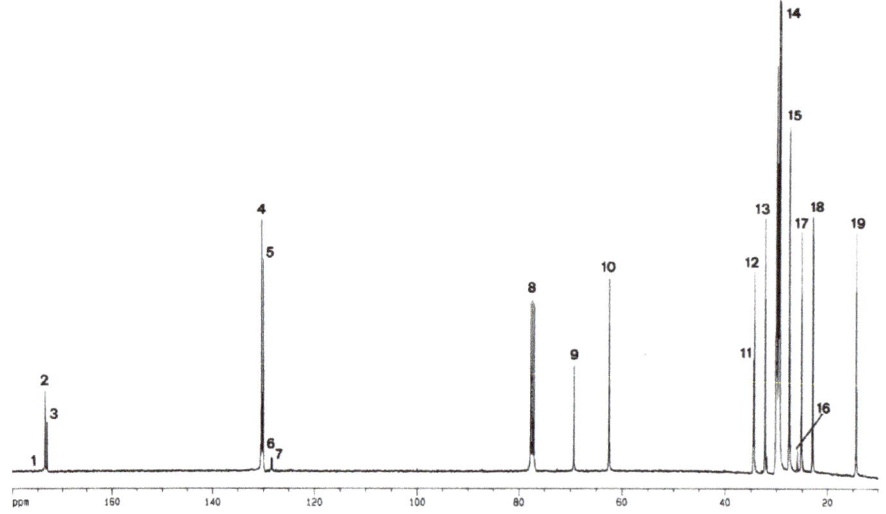

Figure 4. ^{13}C NMR spectrum of a virgin olive oil in $CDCl_3$. Target peaks 1 to 3 in the carbonyl region correspond respectively to: C1 of free fatty acids (174–176 ppm), C1 in 1,3-*sn* position of triacylglycerols (173.26 ppm) and C1 in *sn*-2 position of triacylglycerols (172.81 ppm). For full assignment, see Table 2 of [52]. The spectrum was acquired with broadband proton decoupling on a Bruker AM 400 MHz spectrometer at 303 K, using 16 K data points, with an acquisition time of 0.37 s, a 45° flip angle and a relaxation delay of 5 s. Reprinted from *Magnetic Resonance in Chemistry*, Vol 35, R. Sacchi, F. Addeo, L. Paolillo, "^{1}H and ^{13}C NMR of Virgin Olive Oil. An Overview" (pages S133–S145), copyright (1997), with permission from John Wiley and Sons.

As an alternative method, the region corresponding to aliphatic carbons has been also used for the detection and quantitative determination of FFA in palm oil by ^{13}C NMR spectroscopy [53]. Indeed, the peak for the C3 of an FFA is located at 24.67 ppm, which is 0.16 ppm lower than the peak for C3 of 1,3-acyl groups of the TG in palm oil (Figure 5). In this way, it was possible to easily detect and quantify the composition of FFA in palm oil, as the total integral of the signal at 24.69 ppm represents one FFA chain, whereas the complex band for C3 centred at 24.85 ppm represents three acyl chains of the TG in the oil [53]. The assignment was confirmed by the T_1 value of 0.90 s, measured in $CDCl_3$ solution at 301 K for the C3 in the FFA. Note that the shorter T_1 relaxation times of aliphatic carbons compared to carbonyl carbons makes this method advantageous for the quantitative measurements. Even if the method was developed for palm oil, it can be, in principle, extended to any other vegetable oil.

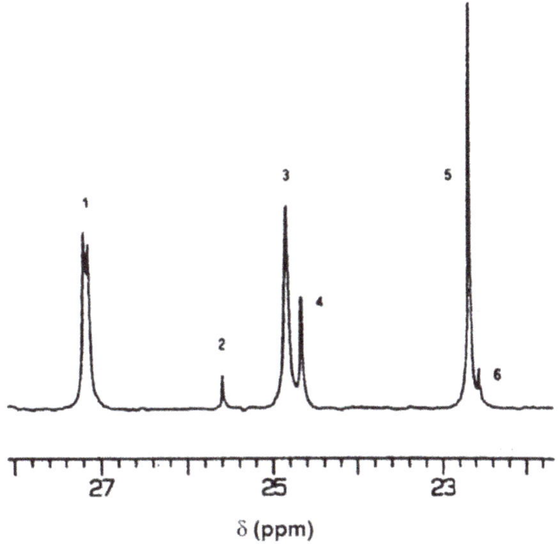

Figure 5. ^{13}C NMR spectrum of the aliphatic carbons in palm oil containing FFA, in $CDCl_3$ solution (concentration 1:3, *v*/*v*). Peak 1: allylic carbons of oleic and linoleic acyls in both triglyceride (TG) and FFA. Peak 2: C11 of linoleic acid. Peak 3: C3 of all acyl chains in TG. Peak 4: C3 of FFA. Peak 5: C15 of palmitic, C17 of stearic, and C17 of oleic acids in TG. Peak 6: C17 of linoleic acid in TG. The spectrum was acquired with broadband proton decoupling on a JEOL FX100 spectrometer at 301 K, using 6000 scans, with an acquisition time of 1.517 s, a 25° flip angle and a relaxation delay of 3 s. Reprinted from the *Journal of the American Oil Chemists' Society*, Vol 77, S. Ng, "Quantitative Analysis of Partial Acylglycerols and Free Fatty Acids in Palm Oil by ^{13}C Nuclear Magnetic Resonance Spectroscopy" (pages 749–755), copyright (2000), with permission from John Wiley and Sons.

2.3. ^{31}P NMR

Edible, waste and recycled vegetable oils are, in principle, phosphorus-free. This is because the crude vegetable oil is subjected to a degumming process, aimed to remove gums, waxes and phospholipids [56]. Nevertheless, ^{31}P NMR spectroscopy has been used as a complementary or alternative tool to ^{1}H NMR and ^{13}C NMR, especially when the analysis is complicated by strong signal overlap in ^{1}H NMR spectra or long relaxation times in ^{13}C NMR spectra [41]. In vegetable oils, alcohols or carboxylic acids constitute many of the minor components of interest. Hence, the strategy is to convert these species into derivatives with the ^{31}P NMR active nucleus. Basically, a phosphitylating reagent, 2-chloro-4,4,5,5-tetramethyldioxaphospholane, is used, which reacts quickly and quantitatively with the hydroxyl and/or carboxyl groups under mild conditions. In detail, according to the protocol originally reported by Dais et al. [57], in order to obtain the oil samples for the ^{31}P NMR analysis, a stock solution (10 mL) is first prepared mixing pyridine and $CDCl_3$ in a 1.6:1.0 volume ratio and adding 0.6 mg of chromium acetylacetonate [$Cr(acac)_3$] and 13.5 mg of cyclohexanol. Then, 150 mg of the oil sample is weighted directly in a 5 mm NMR tube, before adding 0.4 mL of the stock solution and 5–30 μL of the phosphitylating reagent. The mixture is left to react for 20–30 min at room temperature. In this way, the labile hydrogen atoms of oil constituents bearing hydroxyl and/or carboxyl groups are derivatized, then ^{31}P chemical shifts are used to identify the labile centres [57].

Given the high gyromagnetic ratio (10.839×10^7 rad T^{-1} s^{-1}), the wide range of chemical shifts ensuring a good separation of signals located in different environments, the 100% natural abundance of the ^{31}P nucleus, and its high sensitivity (only ~15 times less than that of ^1H), ^{31}P NMR experiments represent a reliable analytical tool able to detect very low concentrations [2,57]. Moreover, the resonance frequency of ^{31}P is very sensitive to the chemical surroundings within the molecule. Furthermore, by introducing, in the reaction mixture, an internal standard, one can obtain the concentration of the product, and then the concentration of the original compound bearing the functional group [57]. In the original method by Dais and co-workers [57], cyclohexanol is used as internal standard. Alternatively, ^{31}P NMR chemical shifts of the phospholane derivatives can be referenced against an internal standard solution of benzoic acid in $CDCl_3$ [4]. The chemical shift of benzoic acid is, in turn, determined using, as an external standard, phosphoric acid at 85% (0.0 ppm). The solvent is sealed in a capillary tube and inserted into the NMR tube. However, it should be noted that to obtain quantitative ^{31}P spectra it is necessary to use the inverse gated decoupling technique together with a spin relaxation agent such as the paramagnetic chemical shift reagent [$Cr(acac)_3$], which lowers the relaxation times of ^{31}P nuclei, and hence shortens the duration of the measurements significantly [2,57,58].

Following this methodology, minor constituents of olive oil, including FFA, monoglycerides (MGs), diglycerides (DGs) and sterols, have been derivatized into phosphorus-containing dioxaphospholane compounds, in order to be identified and quantified by ^{31}P NMR. The acidity can be calculated by integrating the signal of the phosphitylated free fatty acids centred at 134.8 ppm (Figure 6) [59].

It has been demonstrated that, when ^{31}P NMR is applied to the measurement of various major and minor constituents in olive oil, it gives comparable results to conventional analytical methods. Indeed, 1-MGs, 2-MGs, 1,2-DGs, 1,3-DGs, sterols, polyphenols and FFA present in olive oil samples can be unambiguously detected and reliably quantified [41,57–61]. For instance, in a comparative study, Dais et al. measured the free acidity of 137 olive oil samples by titration, following the European Communities official method, and by ^{31}P NMR [41]. Linear regression analyses based on two different methods showed strong agreement between ^{31}P NMR and conventional methods for free acidity, with a correlation coefficient of 0.994 and 96.4% of the measurements (132 out of 137) within the limits of agreement. Only three measurements (2.2%) were near or at the limits, and only two outliers (1.5%) were detected well outside the limits. In another study performed on coconut oil, Dayrit et al. compared the FFA content estimated by ^{31}P NMR with that obtained by standard titrimetry, obtaining comparable results [4].

The main drawback of the ^{31}P approach is clearly the need for the derivatisation of FFA prior to ^{31}P analysis, and hence the destruction of the analytes. However, the methodology is much faster than the corresponding conventional titrimetric methods [41] and can offer some advantages with respect to ^1H or ^{13}C NMR methods. The higher sensitivity of ^{31}P NMR with respect to ^1H NMR was, for instance, demonstrated in a study of the degradation process of olive oil [62]. Seven olive oil samples were subjected to conventional heating and microwave heating and NMR spectroscopy was applied to the analysis of fresh or degraded olive oils. Despite the fact that the ^1H NMR spectra gave evidence of the formation of secondary products after thermal treatment, the ^{31}P NMR protocol was preferred in terms of sensitivity [62]. The FFA quantification via the corresponding signal at 134.8 ppm indicated that the FFA content did not increase significantly after microwave heating, whereas it increased markedly after conventional heating.

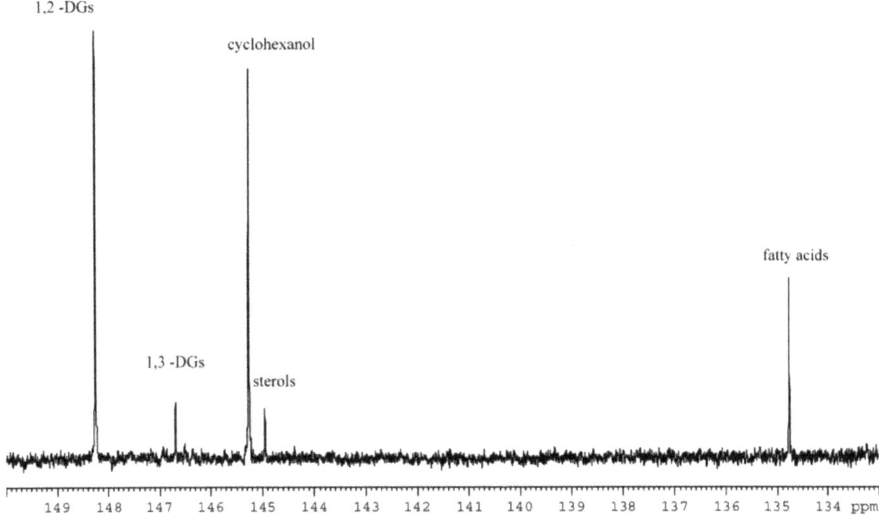

Figure 6. ^{31}P NMR spectrum of virgin olive oil in pyridine/$CDCl_3$, corresponding to the region of phosphitylated total sterols, diglycerides, and free fatty acids. The phosphitylated cyclohexanol is used as internal standard. The spectrum was acquired with inverse-gated decoupling on a Bruker AMX500 spectrometer at room temperature, using 32 scans and 16K data points, with a 90° pulse and a relaxation delay of 30 s. Reprinted with permission from the *Journal of Agricultural and Food Chemistry*, Vol. 51, G. Vigli, A. Philippidis, A. Spyros, P. Dais, "Classification of Edible Oils by Employing ^{31}P and ^{1}H NMR Spectroscopy in Combination with Multivariate Statistical Analysis. A Proposal for the Detection of Seed Oil Adulteration in Virgin Olive Oils" (pages 5715–5722). Copyright (2003) American Chemical Society.

3. Conclusions

Different strategies reported in the literature based on ^{1}H, ^{13}C and ^{31}P NMR spectroscopy for the determination of free fatty acids in vegetable oils have been reviewed and compared. Two ^{1}H NMR methods, based on the integration of the α-carbonyl methylene protons or of the carboxyl proton signal of FFA were reported. The main drawbacks are related to (i) the narrow spectral width and the following risk of signal overlap, (ii) sensitivity issues in the first method, and (iii) the need for some effort in the sample preparation in the second method. In cases of severe overlap in the proton spectrum, ^{13}C methods may be useful, as they rely on the identification and quantification of signals in the carbonyl or the aliphatic region. The main limitation of ^{13}C NMR spectroscopy is the long acquisition time required to obtain a spectrum with a proper signal-to-noise ratio. To solve the problem and shorten the experimental time, it is possible to add relaxation reagents to the samples. An additional method is represented by the ^{31}P NMR of phosphitylated free fatty acids. The need for the derivatisation of FFA prior to ^{31}P analysis represents the main downside of the ^{31}P approach.

All NMR approaches give reliable results in agreement with conventional methods, and can represent, therefore, a non-invasive, non-destructive and quantitative analytical toolbox for the determination of free acidity in vegetable oils, including waste cooking oils.

Typical limitations of all NMR-based techniques are their sensitivity and limits of detection, which could be relevant issues, especially for commercial edible oils or pharmaceutical products. However, impressive progress was made in this respect over the years and, nowadays, the availability of ultra-high magnetic fields and new generation probeheads make it possible to incredibly reduce the noise and push the NMR detection limits in terms of absolute sensitivity.

Author Contributions: Conceptualization, M.E.D.P. and A.M. (Alberto Mannu); resources, A.M. (Alberto Mannu); data curation, M.E.D.P. and A.M. (Alberto Mannu); writing—original draft preparation, M.E.D.P; writing—review and editing, M.E.D.P., A.M. (Alberto Mannu) and A.M. (Andrea Mele); visualization, M.E.D.P.; supervision, M.E.D.P. and A.M. (Andrea Mele). All authors have read and agreed to the published version of the manuscript.

Funding: This research received no external funding.

Acknowledgments: MEDP thanks Politecnico di Milano for her postdoctoral fellowship in the framework of the "MSCA EF Master Class 2018". The Authors thank all of the people involved in the MSCA RISE 2019 project proposal WORLD (873005).

Conflicts of Interest: The authors declare no conflict of interest.

References

1. Skiera, C.; Steliopoulos, P.; Kuballa, T.; Diehl, B.; Holzgrabe, U. Determination of free fatty acids in pharmaceutical lipids by 1H NMR and comparison with the classical acid value. *J. Pharm. Biomed. Anal.* **2014**, *93*, 43–50. [CrossRef] [PubMed]
2. Alexandri, E.; Ahmed, R.; Siddiqui, H.; Choudhary, M.I.; Tsiafoulis, C.G.; Gerothanassis, I.P. High resolution NMR spectroscopy as a structural and analytical tool for unsaturated lipids in solution. *Molecules* **2017**, *22*, 1663. [CrossRef] [PubMed]
3. Skiera, C.; Steliopoulos, P.; Kuballa, T.; Holzgrabe, U.; Diehl, B. Determination of free fatty acids in edible oils by 1H NMR spectroscopy. *Lipid Technol.* **2012**, *24*, 279–281. [CrossRef]
4. Dayrit, F.M.; Buenafe, O.E.M.; Chainani, E.T.; De Vera, I.M.S. Analysis of monoglycerides, diglycerides, sterols, and free fatty acids in coconut (*Cocos nucifera* L) oil by 31P NMR spectroscopy. *J. Agric. Food Chem.* **2008**, *56*, 5765–5769. [CrossRef] [PubMed]
5. Banani, R.; Youssef, S.; Bezzarga, M.; Abderrabba, M. Waste frying oil with high levels of free fatty acids as one of the prominent sources of biodiesel production. *J. Mater. Environ. Sci.* **2015**, *6*, 1178–1185.
6. Mannu, A.; Ferro, M.; Di Pietro, M.E.; Mele, A. Innovative applications of waste cooking oil as raw material. *Sci. Prog.* **2019**, *102*, 153–160. [CrossRef]
7. Mannu, A.; Vlahopoulou, G.; Sireus, V.; Petretto, G.L.; Mulas, G.; Garroni, S. Bentonite as a refining agent in waste cooking oils recycling: Flash point, density and color evaluation. *Nat. Prod. Commun.* **2018**, *13*, 613–616. [CrossRef]
8. Bautista, L.F.; Vicente, G.; Rodríguez, R.; Pacheco, M. Optimisation of FAME production from waste cooking oil for biodiesel use. *Biomass Bioenergy* **2009**, *33*, 862–872. [CrossRef]
9. Bockisch, M. (Ed.) *Fats and Oils Handbook*; AOCS Press: Urbana, IL, USA, 1998; ISBN 9780981893600.
10. Lanser, A.C.; List, G.R.; Holloway, R.K.; Mounts, T.L. FTIR estimation of free fatty acid content in crude oils extracted from damaged soybeans. *J. Am. Oil Chem. Soc.* **1991**, *68*, 448–449. [CrossRef]
11. Li, G.; You, J.; Suo, Y.; Song, C.; Sun, Z.; Xia, L.; Zhao, X.; Shi, J. A developed pre-column derivatization method for the determination of free fatty acids in edible oils by reversed-phase HPLC with fluorescence detection and its application to *Lycium barbarum* seed oil. *Food Chem.* **2011**, *125*, 1365–1372. [CrossRef]
12. Mahesar, S.A.; Sherazi, S.T.H.; Khaskheli, A.R.; Kandhro, A.A.; Uddin, S. Analytical approaches for the assessment of free fatty acids in oils and fats. *Anal. Methods* **2014**, *6*, 4956–4963. [CrossRef]
13. Mannu, A.; Ferro, M.; Dugoni, G.C.; Panzeri, W.; Petretto, G.L.; Urgeghe, P.; Mele, A. Improving the recycling technology of waste cooking oils: Chemical fingerprint as tool for non-biodiesel application. *Waste Manag.* **2019**, *96*, 1–8. [CrossRef] [PubMed]
14. Mannu, A.; Ferro, M.; Colombo Dugoni, G.; Garroni, S.; Taras, A.; Mele, A. Response Surface Analysis of density and flash point in recycled Waste Cooking Oils. *Chem. Data Collect.* **2020**, *25*, 100329. [CrossRef]
15. Vlahopoulou, G.; Petretto, G.L.; Garroni, S.; Piga, C.; Mannu, A. Variation of density and flash point in acid degummed waste cooking oil. *J. Food Process. Preserv.* **2018**, *42*, 1–6. [CrossRef]
16. Mannu, A.; Vlahopoulou, G.; Urgeghe, P.; Ferro, M.; Del Caro, A.; Taras, A.; Garroni, S.; Rourke, J.P.; Cabizza, R.; Petretto, G.L. Variation of the chemical composition ofwaste cooking oils upon bentonite filtration. *Resources* **2019**, *8*, 108. [CrossRef]
17. Berezin, O.Y.; Tur'yan, Y.I.; Kuselman, I.; Shenhar, A. Alternative methods for titratable acidity determination. *Talanta* **1995**, *42*, 507–517. [CrossRef]

18. Tur'yan, Y.I.; Berezin, O.Y.; Kuselman, I.; Shenhar, A. pH-metric determination of acid values in vegetable oils without titration. *J. Am. Oil Chem. Soc.* **1996**, *73*, 295–301. [CrossRef]
19. Lau, H.L.N.; Puah, C.W.; Choo, Y.M.; Ma, A.N.; Chuah, C.H. Simultaneous quantification of free fatty acids, free sterols, squalene, and acylglycerol molecular species in palm oil by high-temperature gas chromatography-flame ionization detection. *Lipids* **2005**, *40*, 523–528.
20. Wan, P.J.; Dowd, M.K.; Thomas, A.E.; Butler, B.H. Trimethylsilyl derivatization/gas chromatography as a method to determine the free fatty acid content of vegetable oils. *J. Am. Oil Chem. Soc.* **2007**, *84*, 701–708. [CrossRef]
21. Kanya, T.C.S.; Rao, L.J.; Sastry, M.C.S. Characterization of wax esters, free fatty alcohols and free fatty acids of crude wax from sunflower seed oil refineries. *Food Chem.* **2007**, *101*, 1552–1557. [CrossRef]
22. Bazina, N.; He, J. Analysis of fatty acid profiles of free fatty acids generated in deep-frying process. *J. Food Sci. Technol.* **2018**, *55*, 3085–3092. [CrossRef]
23. Tarvainen, M.; Suomela, J.P.; Kallio, H. Ultra high performance liquid chromatography-mass spectrometric analysis of oxidized free fatty acids and acylglycerols. *Eur. J. Lipid Sci. Technol.* **2011**, *113*, 409–422. [CrossRef]
24. Yu, X.; Van De Voort, F.R.; Sedman, J.; Gao, J.M. A new direct Fourier transform infrared analysis of free fatty acids in edible oils using spectral reconstitution. *Anal. Bioanal. Chem.* **2011**, *401*, 315–324. [CrossRef] [PubMed]
25. Che Man, Y.B.; Setiowaty, G. Application of Fourier transform infrared spectroscopy to determine free fatty acid contents in palm olein. *Food Chem.* **1999**, *66*, 109–114. [CrossRef]
26. Moschner, C.R.; Biskupek-Korell, B. Estimating the content of free fatty acids in high-oleic sunflower seeds by near-infrared spectroscopy. *Eur. J. Lipid Sci. Technol.* **2006**, *108*, 606–613. [CrossRef]
27. Gerde, J.A.; Hardy, C.L.; Hurburgh, C.R.; White, P.J. Rapid determination of degradation in frying oils with near-infrared spectroscopy. *J. Am. Oil Chem. Soc.* **2007**, *84*, 519–522. [CrossRef]
28. Novak, M. Colorimetric Ultramicro Method for the Determination of Free Fatty Acids. *J. Lipid Res.* **1965**, *6*, 431–433.
29. Gyrik, M.; Ajtony, Z.; Dóka, O.; Alebic-Juretić, A.; Bicanic, D.; Koudijs, A. Determination of free fatty acids in cooking oil: Traditional spectrophotometry and optothermal window assay. *Instrum. Sci. Technol.* **2006**, *34*, 119–128. [CrossRef]
30. Lowry, R.R.; Tinsley, I.J. Rapid Colorimetric Determination of Free Fatty Acids. *J. Am. Oil Chem. Soc.* **1976**, *53*, 470–472. [CrossRef]
31. Mariotti, E.; Mascini, M. Determination of extra virgin olive oil acidity by FIA-titration. *Food Chem.* **2001**, *73*, 235–238. [CrossRef]
32. Saad, B.; Ling, C.W.; Jab, M.S.; Lim, B.P.; Mohamad Ali, A.S.; Wai, W.T.; Saleh, M.I. Determination of free fatty acids in palm oil samples using non-aqueous flow injection titrimetric method. *Food Chem.* **2007**, *102*, 1407–1414. [CrossRef]
33. Zhi, Z.L.; Ríos, A.; Valcárcel, M. An automated flow-reversal injection/liquid-liquid extraction approach to the direct determination of total free fatty acids in olive oils. *Anal. Chim. Acta* **1996**, *318*, 187–194. [CrossRef]
34. Lankhorst, P.P.; Chang, A.N. The Application of NMR in Compositional and Quantitative Analysis of Oils and Lipids. In *Modern Magnetic Resonance*; Webb, G.A., Ed.; Springer International Publishing AG: Cham, Switzerland, 2018; pp. 1743–1764, ISBN 9783319283883.
35. Diehl, B.W.K. Multinuclear high-resolution nuclear magnetic resonance spectroscopy. In *Lipid Analysis in Oils and Fats*; Hamilton, R.J., Ed.; Blackie Academic & Professional, Thomson Science: London, UK, 1997; pp. 87–135, ISBN 9788578110796.
36. Lie Ken Jie, M.S.F.; Mustafa, J. High-resolution nuclear magnetic resonance spectroscopy—Applications to fatty acids and triacylglycerols. *Lipids* **1997**, *32*, 1019–1034. [CrossRef] [PubMed]
37. Gunstone, F.D. High Resolution 13C NMR. A Technique for the Study of Lipid Structure and Composition. *Progr. Lipid Res.* **1994**, *33*, 19–28. [CrossRef]
38. Gunstone, F.D.; Shukla, V.K.S. NMR of lipids. *Annu. Rep. NMR Spectrosc.* **1995**, *31*, 219–237.
39. Miyake, Y.; Yokomizo, K.; Matsuzaki, N. Rapid Determination of Iodine Value by 1H Nuclear Magnetic Resonance Spectroscopy. *J. Am. Oil Chem. Soc.* **1998**, *75*, 15–19. [CrossRef]
40. Guillén, M.D.; Ruiz, A. Rapid simultaneous determination by proton NMR of unsaturation and composition of acyl groups in vegetable oils. *Eur. J. Lipid Sci. Technol.* **2003**, *105*, 688–696. [CrossRef]

41. Dais, P.; Spyros, A.; Christophoridou, S.; Hatzakis, E.; Fragaki, G.; Agiomyrgianaki, A.; Salivaras, E.; Siragakis, G.; Daskalaki, D.; Tasioula-Margari, M.; et al. Comparison of analytical methodologies based on 1H and 31P NMR spectroscopy with conventional methods of analysis for the determination of some olive oil constituents. *J. Agric. Food Chem.* **2007**, *55*, 577–584. [CrossRef]
42. Fauhl, C.; Reniero, F.; Guillou, C. 1H NMR as a tool for the analysis of mixtures of virgin olive oil with oils of different botanical origin. *Magn. Reson. Chem.* **2000**, *38*, 436–443. [CrossRef]
43. Guillén, M.D.; Ruiz, A. High resolution 1H nuclear magnetic resonance in the study of edible oils and fats. *Trends Food Sci. Technol.* **2001**, *12*, 328–338. [CrossRef]
44. The Human Metabolome Database. Available online: https://hmdb.ca/. (accessed on 27 March 2020).
45. Wishart, D.S.; Feunang, Y.D.; Marcu, A.; Guo, A.C.; Liang, K.; Vázquez-Fresno, R.; Sajed, T.; Johnson, D.; Li, C.; Karu, N.; et al. HMDB 4.0: The human metabolome database for 2018. *Nucleic Acids Res.* **2018**, *46*, D608–D617. [CrossRef] [PubMed]
46. Satyarthi, J.K.; Srinivas, D.; Ratnasamy, P. Estimation of free fatty acid content in oils, fats, and biodiesel by 1H NMR spectroscopy. *Energy Fuels* **2009**, *23*, 2273–2277. [CrossRef]
47. Kumar, R.; Bansal, V. Estimation of Glycerides and Free Fatty Acid in Oils Extracted From Various Seeds from the Indian Region by NMR Spectroscopy. *J. Am. Oil Chem. Soc.* **2011**, *88*, 1675–1685. [CrossRef]
48. Annarao, S.; Sidhu, O.P.; Roy, R.; Tuli, R.; Khetrapal, C.L. Lipid profiling of developing *Jatropha curcas* L. seeds using 1H NMR spectroscopy. *Bioresour. Technol.* **2008**, *99*, 9032–9035. [CrossRef]
49. Nieva-Echevarría, B.; Goicoechea, E.; Manzanos, M.J.; Guillén, M.D. A method based on 1H NMR spectral data useful to evaluate the hydrolysis level in complex lipid mixtures. *Food Res. Int.* **2014**, *66*, 379–387. [CrossRef]
50. Kumar, R.; Das, N. Seed oil of *Jatropha curcas* L. germplasm: Analysis of oil quality and fatty acid composition. *Ind. Crops Prod.* **2018**, *124*, 663–668. [CrossRef]
51. San Martín, E.; Avenoza, A.; Peregrina, J.M.; Busto, J.H. Solvent-based strategy improves the direct determination of key parameters in edible fats and oils by 1 H NMR. *J. Sci. Food Agric.* **2020**, *100*, 1726–1734. [CrossRef] [PubMed]
52. Sacchi, R.; Addeo, F.; Paolillo, L. 1H and 13C NMR of Virgin Olive Oil. An Overview. *Magn. Reson. Chem.* **1997**, *35*, 133–145. [CrossRef]
53. Ng, S. Quantitative analysis of partial acylglycerols and free fatty acids in palm oil by 13C nuclear magnetic resonance spectroscopy. *J. Am. Oil Chem. Soc.* **2000**, *77*, 749–755. [CrossRef]
54. Sacchi, R.; Medina, J.I.; Aubourg, S.P.; Giudicianni, I.; Paolillo, L.; Addeo, F. Quantitative High-Resolution 13C NMR Analysis of Lipids Extracted from the White Muscle of Atlantic Tuna (*Thunnus alalunga*). *J. Agric. Food Chem.* **1993**, *41*, 1247–1253. [CrossRef]
55. Scano, P.; Rosa, A.; Marincola, F.C.; Locci, E.; Melis, M.P.; Dess, M.A.; Lai, A. C NMR, GC and HPLC characterization of lipid components of the salted and dried mullet (*Mugil cephalus*) roe " bottarga". *Chem. Phys. Lipids* **2008**, *151*, 69–76. [CrossRef] [PubMed]
56. O'Brien, R.D. Soybean Oil Purification. In *Soybeans. Chemistry, Production, Processing, and Utilization*; Johnson, L.A., White, P.J., Galloway, R., Eds.; Academic Press: Cambridge, MA, USA; AOCS Press: Urbana, IL, USA, 2008; pp. 377–408, ISBN 978-1-893997-64-6.
57. Spyros, A.; Dais, P. Application of 31P NMR spectroscopy in food analysis. 1. Quantitative determination of the mono- and diglyceride composition of olive oils. *J. Agric. Food Chem.* **2000**, *48*, 802–805. [CrossRef] [PubMed]
58. Fronimaki, P.; Spyros, A.; Christophoridou, S.; Dais, P. Determination of the diglyceride content in Greek virgin olive oils and some commercial olive oils by employing 31P NMR spectroscopy. *J. Agric. Food Chem.* **2002**, *50*, 2207–2213. [CrossRef] [PubMed]
59. Vigli, G.; Philippidis, A.; Spyros, A.; Dais, P. Classification of edible oils by employing 31P and 1H NMR spectroscopy in combination with multivariate statistical analysis. A proposal for the detection of seed oil adulteration in virgin olive oils. *J. Agric. Food Chem.* **2003**, *51*, 5715–5722. [CrossRef]
60. Christophoridou, S.; Dais, P. Novel approach to the detection and quantification of phenolic compounds in olive oil based on 31P nuclear magnetic resonance spectroscopy. *J. Agric. Food Chem.* **2006**, *54*, 656–664. [CrossRef]

61. Christophoridou, S.; Spyros, A.; Dais, P. 31P nuclear magnetic resonance spectroscopy of polyphenol-containing olive oil model compounds. *Phosphorus Sulfur Silicon Relat. Elem.* **2001**, *170*, 139–157. [CrossRef]
62. Lucas-Torres, C.; Pérez, Á.; Cabañas, B.; Moreno, A. Study by 31 P NMR spectroscopy of the triacylglycerol degradation processes in olive oil with different heat-transfer mechanisms. *Food Chem.* **2014**, *165*, 21–28. [CrossRef]

Review

Available Technologies and Materials for Waste Cooking Oil Recycling

Alberto Mannu [1,*], Sebastiano Garroni [2,*], Jesus Ibanez Porras [3] and Andrea Mele [1,4]

1 Department of Chemistry, Materials and Chemical Engineering "G. Natta", Politecnico di Milano, Piazza L. da Vinci 32, 20133 Milano, Italy; andrea.mele@polimi.it
2 Department of Chemistry and Pharmacy, via Vienna 2, University of Sassari, 07100 Sassari, Italy
3 International Research Centre in Critical Raw Materials (ICCRAM), University of Burgos, Plaza Misael Bañuelos s/n, 09001 Burgos, Spain; jesusibanez@ubu.es
4 CNR-ICRM Istituto di Chimica del Riconoscimento Molecolare, 'U.O.S. Milano Politecnico', 20131 Milano, Italy
* Correspondence: alberto.mannu@polimi.it (A.M.); sgarroni@uniss.it (S.G.)

Received: 14 February 2020; Accepted: 12 March 2020; Published: 22 March 2020

Abstract: Recently, the interest in converting waste cooking oils (WCOs) to raw materials has grown exponentially. The driving force of such a trend is mainly represented by the increasing number of WCO applications, combined with the definition, in many countries, of new regulations on waste management. From an industrial perspective, the simple chemical composition of WCOs make them suitable as valuable chemical building blocks, in fuel, materials, and lubricant productions. The sustainability of such applications is sprightly related to proper recycling procedures. In this context, the development of new recycling processes, as well as the optimization of the existing ones, represents a priority for applied chemistry, chemical engineering, and material science. With the aim of providing useful updates to the scientific community involved in vegetable oil processing, the current available technologies for WCO recycling are herein reported, described, and discussed. In detail, two main types of WCO treatments will be considered: chemical transformations, to exploit the chemical functional groups present in the waste for the synthesis of added value products, and physical treatments as extraction, filtration, and distillation procedures. The first part, regarding chemical synthesis, will be connected mostly to the production of fuels. The second part, concerning physical treatments, will focus on bio-lubricant production. Moreover, during the description of filtering procedures, a special focus will be given to the development and applicability of new materials and technologies for WCO treatments.

Keywords: waste cooking oil; biolubricant; biodiesel; recycling; vegetable oil filtration; vegetable oil degumming

1. Introduction

Amongst the many issues currently faced by the scientific community, the interest toward the optimization of resources, while reducing the environmental impact of new and existing production processes, are of major interest. The amount of research focused on these transversal topics are growing, exponentially, in many research fields [1]. As a general approach, to develop such ambitious research, many scientists have explored routes for the employment of waste as raw material in new productions, as well as for the reconversion of existing processes. This is not an easy road, as industry requires to reach such targets in a very short time. The most demanding step in this scientific activity is represented by the passage from a technology readiness level (TRL) of 3 (analytical and experimental critical function and/or characteristic proof of concept) to 6 (system/subsystem model or prototype demonstration in a relevant environment) [2]. In fact, the transfer of knowledge is often hampered by the limits of the current available technologies. In this context, the development of improved technologies and materials for the transformation of waste-based raw materials into valuable products represents a primary target to unlock many transformations that exist only at design level or laboratory scale [3].

Within the restricted list of potential waste-based raw materials—which can lead, if properly processed, to a multiple typology of products—waste cooking oil (WCO) stands out [4].

WCO pertains to the family of used vegetable oils (UVOs), which are considered waste that is dangerous for the environment [5]. WCOs are the main representatives of such family, as most of the collected UVOs arise from kitchens and catering industries [6]. As frying food represents the main worldwide employed cooking method, WCOs are geographically diffused and produced everywhere in large amounts. The yearly overall production of used vegetable oils exceeds 190 million metric tons, with the European Union (EU) contributing about 1 million tons/year [7,8].

A proper handling of WCOs is mandatory in many countries (and currently the subject of political discussion in others) [9–11]. The issues regarding the treatment of WCOs are mainly two: the disposal-collection strategy, and the reconversion of the waste [12]. Disposal and collection systems will not be discussed in the present review paper, as a good amount of information about this specific topic can be found elsewhere [4,13].

Regarding the reconversion of WCOs, they can be used as main raw materials in many industrial processes, such as bio-lubricant [14] or fuel [15–17] production, or as additives for asphalts [18] and animal feed [19]. Other possible employments of WCOs are strictly related to their chemical compositions. WCOs are basically a mixture of triglycerides and fatty acids, contaminated by some derivatives during the frying process, as free fatty acids (FFA), heterocycles, Maillard reaction products, and metal traces originated from pads and food leaching [20,21]. The specific composition of WCOs can also be exploited as useful sources of chemicals for the production of bio-plasticizers, syngas, and sorbents for volatile organic compounds (VOCs) [22,23].

In the present review paper, some technological advancements related to the employment of WCOs as raw materials will be described and discussed. A special focus will be given to the period that ranges from 2015 to early 2020.

2. Technologies and Materials for the Chemical Treatment of WCOs

In this section, chemical treatments of WCOs that are amenable to practical application will be discussed.

WCOs are mixtures of long chain fatty acids (mainly linoleic, linolenic, and oleic), in form of tri- di- and mono-glycerides, and a variable percentage of free fatty acids (FFA); they represent a platform of raw materials for many industries [24].

The chemical composition of WCOs is quite similar to one of the parent edible oils, and differs from the former in terms of decomposition and leaching products. During the frying process, a portion of triglycerides, of the ester moiety, break down. The degree of such degradation depends on the number of frying cycles, frying time, temperature, and the specific vegetable oil [25]. Moreover, during deep frying, many volatile compounds are generated as a consequence of a combination between high temperature and oxygen, which promotes oxidation processes, and other transformations (e.g., the Maillard process) [26,27]. In addition, food and tool exposure during frying promote leaching, enriching the oil composition with metal traces, spices, and other organic molecules [28]. Analysis of the volatile fraction of WCOs reveal a complex mixture of chemicals, which include aldehydes, alcohols, dienes, and heterocycles. In particular, samples of commercial sunflower oil were analyzed by Mannu et al. [29], prior to and after several cycles of frying. Many chemicals were detected in the samples subject to frying, such as hexanal, heptanal, limonene, furan, 2-pentyl-, nonanal, 1-octen-3-ol, furfural, cyclohexanol-dimethyl-2, benzaldehyde, 2-nonenal, 2-furan-methanol, 2-decenal, 2-undecenal, and 2,4-decadienal.

The changes in the chemical composition of oil during the frying process can be related to food and tool contamination and decomposition. Nevertheless, the relative amount of impurities generated during the cooking process is not elevated. It is possible to estimate, qualitatively, the chemical composition of WCOs through Nuclear Magnetic Resonance (NMR) analysis. ^{1}H NMR spectra show a low amount of contaminants (less than 5%), confirming that WCOs are very similar, in terms of overall chemical composition, to the parent edible oils [29].

If, from one side, the low degree of contamination and degradation is enough to declare WCOs unfit for food applications [30,31], from the other side, it allows to set-up a (not particularly difficult) recycling process [32].

Typically, the treatment of collected WCOs involve a first gross filtration aimed to remove solid materials dispersed in the oil. This is followed by the direct introduction of crude material as raw material in the production, without the need of specific decontamination/transformation steps. Then, depending to the specific application, the WCO-raw material can be subject to different kinds of transformations.

A very diffuse procedure consists of exploiting the chemical composition of WCOs to generate esters in basic media, and in the presence of alcohol (esterification). Common esters produced by this route are the fatty acid methyl esters (FAMEs), derived by reaction with methanol, and fatty acid ethyl esters (FAEEs), obtained through esterification with ethanol. The industrial destination of FAMEs and FAEEs are usually employed in biodiesel production [33]. In that case, FAEEs show some advantages with respect to FAME as enhanced fuel properties, in terms of stability towards oxidation and superior lubricant power [34]. Regarding fuel application, the assessment of the acidity of the raw material is crucial. In fact, if the free acidity (FA) results higher than 2.5%–3%, the raw material is not suitable for biodiesel, and is subject to previous procedures in order to reduce the amount of FA. This parameter represents one of the main economic indicators for collectors, because it determines the selling price when WCOs are delivered to biodiesel facilities [35]. For companies involved in collecting and delivering large amounts of WCO, the cost for reducing the amount of FA can be elevated. However, on a small scale, collectors can blend WCOs with different acidity until it reaches the appropriate level of FA required by the producers. The possibility to easily bypassing the issue on a small scale allows the collectors to focus more on the optimization of the recovery step, enhancing their competitiveness. In such conditions, the biggest cost is represented by the acquisition of a proper warehouse for WCO storage, which need to be blended. This last aspect can double the rental price and increase the energy consumption to maintain the oil in adequate conditions.

The acidity index of WCOs is also chemically relevant, as it influences the rate of the collateral hydrolysis of the triacylglycerols, which occur during the basic esterification (saponification).

Saponification gives rise to a relevant issue for industries: the formation of emulsions—they are difficult to reduce on a big scale (in biodiesel plants, for example) [36]. The formation of soap is also promoted by the presence of water in WCOs. For this reason, WCOs are often subject to water removing procedures—usually based on decantation—prior to being sold to biodiesel producers.

On the other side, WCOs that are destined to be processed in soap facilities do not have such pH limitation, as the raw materials are processed by a harsh basic treatment. This subject was recently described in a review article by Felix and coworkers [37].

A relevant amount of research has focused on developing routes to avoid collateral saponification during methyl ester synthesis from WCOs. Fereidooni et al. recently reported an electrolytic process for the esterification of WCOs in the presence of MeOH and KOH to obtain FAMEs. In such conditions, the collateral saponification does not occur even in the presence of consistent amounts of water [38].

Moreover, esterification of WCOs do not generate negligible amounts of glycerol, which can promote a collateral transesterification reaction. In order to avoid undesired transesterification, and considering the commercial value of glycerol, this is usually recovered and exploited in parallel processes [39].

As a matter of fact, acidic esterification of WCOs is also possible, and it can be employed for the batch of raw oil, showing elevated values of acidity; thus, not being suitable for biodiesel [40]. Acid-catalyzed esterification of WCOs to FAME was achieved by Quintain and coworkers [41] by employing a graphene oxide-based catalyst under microwave irradiation.

A general overview of the main reactivity of WCOs in different conditions is reported in Scheme 1.

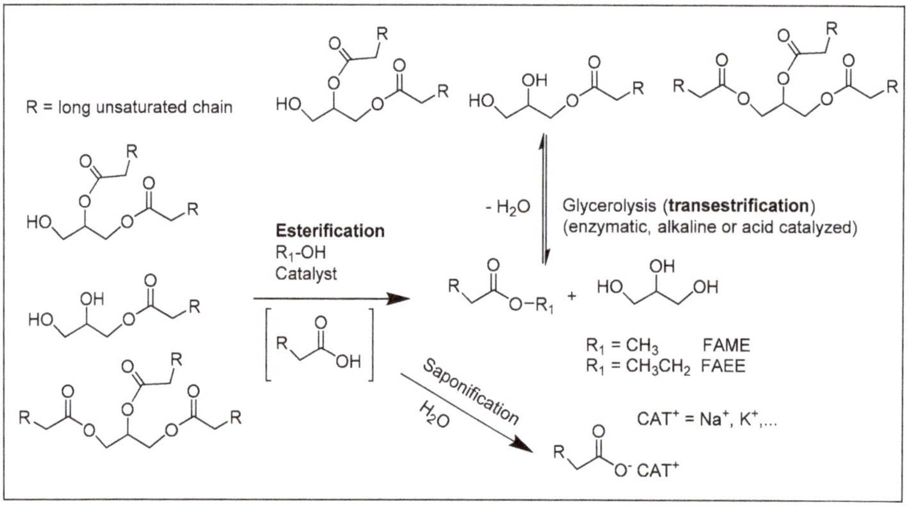

Scheme 1. Esterification of waste cooking oils (WCOs) and related processes.

An alternative alcohol to methanol and ethanol for esterification of WCOs is the glycerol. In that case, the reaction is known as glycerolysis, and in the case of WCOs, formally regenerates, in part, the original composition of the oil (Scheme 1) [42].

The glycerolysis can be exploited for increasing the added value of the starting materials (WCOs) by transforming the free fatty acids (FFA) contained in the waste into added-value chemicals. Mono- and di-glycerides are used as surfactants [43] or as emulsifiers in the food, cosmetic, and pharmaceutical industry [44], while tri-glycerides are employed as additives in bio-diesel production [45,46].

Regarding the effectiveness of glycerolysis, the main issue is represented by the low miscibility of glycerol and waste oil. The problem has been faced for decades; optimized protocols, which usually employ high temperature conditions, have been proposed. Recently, advances toward milder WCO glycerolysis have been reported by Supradan and coworkers [47], who proposed the use of hexane as co-solvent for undertaking mass transfer problems generated by the low solubility of glycerol. Alternatively, Mazubert et al. [48] proposed to perform the glycerolysis of FFA in a reactor system.

Catalytic production of bio-diesel has also faced updates in previous years. Abdillah et al. [49] described the combination of activated carbon obtained from palm oil biomass and potassium phosphate tri-basics (K_3PO_4) as a heterogeneous catalyst for fatty acid methyl esters (FAME) production from WCOs.

Moving to the homogeneous phase, Borovinskaya and coworkers [50] reported the catalyzed trans-esterification of WCOs to obtain FAEEs, performed in a micro-/milli-reactor.

The synthesis of WCO-derived esters can also be exploited in the bio-lubricant industry. What is interesting is the case of the trimethylolpropane fatty acid trimethyl ester, which can be used as a lubricant base stock with superior lubricant power, stability, and biodegradability, with respect to other pentaerythritol fatty esters (TFATEs) [51]. These lubricants can be obtained by transesterification between FAME and trimethylolpropane (TMP) [52] or by esterification of FFAs with TMP. A recent contribution to this specific topic has been offered by Sun and coworkers [53], who proposed the production of TFATEs through transesterification of WCO-derived FAME, with TMP mediated by a catalyst based on a mixture between hydrotalcite and potassium carbonate (K_2CO_3). Recycling of the catalyst, combined with waste-based raw material, makes this process economic and low-impacting. An analogue process, which involves a different catalyst (Mg-Al/hydrotalcite) was recently described by Costarosa et al. [54], and then optimized by the multivariate statistic and response surface analysis (RSA). As a result of such studies, for biodiesel applications, basic catalyzed esterification seems to be the preferred methodology, mainly due to the development of specific process conditions that avoid the formation of emulsions. Similar optimization tools were employed by Mannu et al. [55], who reported optimized conditions for the production of bio-lubricants from water treatment of WCOs as result of RSA. The availability of effective tools to optimize the bio-lubricant production by water treatment of WCOs make this last process competitive, with respect to the filtration procedures, which have been historically employed as a main purification step in vegetable oil refining.

3. Technologies and Materials for the Physical Treatment of WCOs

Physical treatment of WCOs is basically aimed at removing undesired products from crude WCOs, and to obtain a regenerated oil enriched in fatty acids. It is possible to consider three physical routes for achieving WCO regeneration:

(1) Separation based on solubility.
(2) Separation through filtration with specific materials.
(3) Separation based on the boiling point (distillation).

3.1. Separation Based on Solubility

Regarding the separation process based on solubility, water is the solvent of choice. In fact, as the oil is immiscible with water, many chemicals contained in WCOs can easily be removed from WCOs by simply washing the crude material. This procedure can also be extended to an industrial environment by employing a metallic tank equipped with a mechanical stirrer. This (very easy) configuration has been known, for many years as "degumming" (for its purification of vegetable oils). Its aim is to remove phospholipids and waxes [56]. Some local industries employ this procedure for the production of bio-lubricants from WCOs, with any further optimization. Nevertheless, recent research has revealed non-negligible room for improvement. In fact, by tuning the pH of the water, as well as working at specific temperatures and with determined stirring times, it is possible to increase the density, the flash point, and the color of the resulting bio-lubricant [57]. Vlahopoulou et al. [57] reported that by changing the degumming conditions, only small changes could be observed in the density values, while the flash point was much more sensitive to the process parameters. In particular, WCOs subject to different water treatments show a flash point number that ranges from 270 °C (pH = 4, 20 °C, 5 h of stirring, and oil/water wt% ratio = 30), to 290 °C (pH = 6, 60 °C, 5 h of stirring, and oil/water wt% ratio = 30). In Table 1, the variation of the flash point of WCOs after degumming at different conditions is reported. The possibility to increase the flash point value by tuning the previous described parameters is of high industrial interest as it can directly influence the quality of the refined bio-lubricant.

As the variation of the physical parameters is the result of a change in the chemical composition, Mannu and coworkers went through the matter and studied the variation of the volatile profile of WCO samples subjected to water degumming (Table 1) [29].

Looking at the data reported in Table 1, it is possible to confirm the formation of several volatile compounds during the life-cycle of the vegetable oil and their partial removal as a result of the water treatment. As expected, working on pH and temperature range, it is possible to selectively remove specific functional groups, as heptanal and limonene, which are removed only in acidic conditions. This methodology presents some drawbacks, as the saponification observed at pH = 9 (which generates undesired emulsions) or the possibility of corrosion issues when the process treatment is performed at pH = 3 [30].

Table 1. Variation of the volatile profile in samples of edible oil **M1**, fried oil **WCO1**, and degummed samples [29].

Entry	Retention Time	M1	WCO1 [1]	pH = 3 and T = 25 °C	pH = 9 and T = 25 °C	pH = 3 and T = 80 °C	pH = 9 and T = 80 °C	Analyte
1	15.489	n.d.	2.7287	1.9981	1.7933	0.8577	1.8302	Hexanal
2	19.009	n.d.	0.7503	0.2685	0.3263	-	0.6493	Heptanal
3	19.573	n.d.	0.7684	-	0.4901	-	0.4494	Limonene
4	20.354	n.d.	1.8623	1.0677	1.2202	-	0.4494	Furan, 2-pentyl-
5	25.183	n.d.	4.0382	3.8689	4.0172	2.8489	2.4767	Nonanal
6	26.086	n.d.	0.8284	0.6770	0.8705	0.4485	0.7353	1-octen-3-ol
7	27.052	n.d.	0.6783	-	0.5719	0.0791	0.1544	Furfural
8	29.158	n.d.	0.5625	0.4577	0.4587	0.1700	0.5569	Cyclohexanol [2]
9	29.475	n.d.	0.4908	0.4230	0.4550	0.2407	0.2646	Benzaldehyde
10	29.692	n.d.	0.6348	0.6042	0.5006	0.4925	0.7101	2-Nonenal
11	32.182	n.d.	-	-	-	-	-	2-Furan-methanol
12	32.423	n.d.	1.4988	1.4236	1.4708	1.3255	1.5066	2-Decenal
13	34.587	n.d.	1.0504	1.0649	1.0810	1.0060	0.6942	2-Undecenal
14	34.79	n.d.	1.2092	1.1551	1.1331	1.0718	1.0759	2,4-Decadienal [2]

[1] Determined by HS-SPME-GC MS [58], the results are expressed as absolute peak area ($\times 10^6$) obtained by total ion current (TIC) chromatogram. [2] Not identified isomer. n.d. stands for not-detected.

3.2. Separation Through Filtration with Specific Materials

Particle sequestration and separation represents the essence of the recycle process and its efficiency depends on the particulate that can be retained, the volume of the filtrate that can flow through the filter, and physicochemical properties of the filtered oil upon the process. Physical and chemical regeneration of waste frying oils can be typically divided into three main technological approaches: membrane treatment, conventional, and activate filtration. Both the aforementioned approaches exploit different solid materials, which characterize performance, cost, and sustainability of the whole recycling process.

In the last two decades, membrane technology application in vegetable oil processing has received increasing attention due to different advantages, including low energy consumption (approximately 50% with respect to the conventional filtration), and retention of nutrient components [59]. Membranes can be defined as a semipermeable barrier, allowing the passage of components (in this case the vegetable oil) and the retention of others (i.e., the undesired products produced after frying at a high temperature). Membrane filtration is a process driven by the pressure applied and the size of by-products. Two main classes of materials are used for producing commercial membranes: (i) polymers consisting of cellulose acetate, polyamide, polysulfone, and polyvinylidene difluoride (PVDF); and (ii) ceramic and metals materials, including alumina and stainless steel. Moreover, ultra- (UF) and nano-filtration membranes, characterized by their specific pore dimensions (<1 μm), have been largely explored to remove, for example, triglycerides and phospholipids during the degumming step [60]. As an example, a membrane based on PVDF can sequestrate over 98% of phospholipid, with excellent resistance to fouling during the degumming process, and high permeate fluxes, making this membrane very promising for industrial application [61,62]. Furthermore, in addition to phospholipids, these specific membranes can remove up to 80% of color from WCOs [63]. Typically, the filtration process, which involves membrane apparatus at lab scale, is carried out under pressure of 2–5 bars, flow velocity of 0.3–0.4 L/m^3, and in a temperature range of 40–60 °C [62]. Ultrafiltration of membrane prepared from polyethersulfone (PES), removed up to 89% of phospholipid leading to a color reduction, expressed in yellow units of 25% [64]. Recently, Onal-Ulusoy and coworkers improved, significantly, PES-based membrane sequestration activity for waste frying oil recycling, modifying their surfaces by radio frequency (RF) plasma treatment [65]. During membrane filtration, triglyceride components permeated preferentially compared to the polar compounds including oxidation and hydrolysis products and polymers. The experimental results showed that PES membranes, surface modified with HMDSO at 75 W for 5 min, retain selectively free fatty acids (FFAs) of waste frying oil, up to 35.3%–40%, and reduced the oil viscosity 9.4% to 12.8%. Finally, the use of membrane technology appears as a promising alternative to the conventional method of waste oil processing, due to its suitability to be applied at different stages of the whole recycle process (degumming, refining, etc.) and its low energy consumption. However, in order to drastically improve their performance, and to overcome some barriers represented by the optimization of the permeate flow rates, a deeper investigation on the materials is necessary.

Conventional filters, used in the filtration apparatus, include a pad of cellulose fibers and metal screens. Cellulose filters, for example, are very cheap if compared to other materials. Recently, Malagon-Rometo et al. [66] developed and standardized a process to treat Colombian waste cooking oils, which removes 79% of the solid particles, employing paper filter (cellulose). However, cellulose pads absorb an important amount of the filtered oil and they can only retain particles with dimensions larger than 1 μm. Furthermore, they present poor mechanical properties and are often damaged after removing particulates adsorbed after filtration, reducing their use for multiple filtration cycles (up to 600 kg of oil). Stainless steel filter screens, on the other hand, can be reused indefinitely, but are more expensive then cellulose pads, and their efficiency separation is limited to particles large than 80 μm. In order to increase filtration efficiency, typically, a powdered filter aid can be used in combination with a conventional filter. Such filter aids are deposited in the main filter surface, increasing the filtration surface area and then enhancing the retention of smaller particles (around 1 μm). At the same time, filter aids also include adsorbents or neutralizing agents to provide active filtration, as a consequence of chemical interaction and/or electrostatic bonding with free fatty acid contaminants. Examples of filter aids include inorganic compounds, such as diatomaceous earth, clays, silicates, and activated carbons. In the past, synthetic calcium and magnesium silicate has been extensively used to reduce FFA and color, respectively [67]. During the last 20 years, new filter aid products have been synthesized, including blends of silicates with magnesium and aluminum oxides, which have been included into the process for treating used cooking oil in frying operations [68,69]. With the same purpose, filters composed by synthetic fibers of polyester, polypropylene, and nylon, have been developed and tested in a very efficient process, able to remove 93%–98% of fatty acids from cooking oil used [70,71].

Currently, it is possible to find on the market some small-scale machines for WCO regeneration. The exploitability of such technological solutions is strongly influenced by local legislation, as in some countries, the recycling of WCO for food purposes is strictly regulated [72]. For example, some companies sell recycling apparatuses to extend the life of frying oil, claiming to reach up to 50% of additional cooking time [73].

Despite the high performing capacity to remove particulates displayed by conventional, powdered, and fiber filters, the retention of smaller, undesired particles (<0.5 μm) remains a not-yet solved problem. These materials are, in fact, characterized by medium-low surface areas (<500 m^2/g) and macroporosity (d > 0.1 μm). For this reason, increasing attention has been addressed, in previous years, to mesoporous and microporous materials with pore size dimensions of 2–50 nm, and very high specific surface areas (>500 m^2/g) to act as sorbent systems for small particles formed in the cooking oil during the frying process. Recently, for example, it was found that aluminum-, zinc- and titanium-containing metal–organic frameworks (Al-, Zn-, and Ti-MoF) improve significantly, with respect to traditional filters, taste, and odor properties of unrefined vegetable oils, due to the binding of free fatty acids and peroxide compounds [74]. Unfortunately, the high cost synthesis and regeneration treatment are main barriers for rapid commercialization of this class of materials, for this specific purpose. Therefore, it is no surprise that the attention from part of the scientific community, even if only recently, has moved to test natural porous materials, such as bentonites [75]. For example, montmorillonite (the dominant component of bentonite), tested for WCO purification, seems to retain specific chemical groups, such as carboxylic acids or double bonds, significantly improving the color quality of the filtered oil [76].

A transfer of knowledge attempt, based on a combination of water treatment and filtration on porous materials, has been made by Mannu and coworkers, by designing a recycling mini-plant aimed to produce bio-lubricants from WCOs (Figure 1) [29].

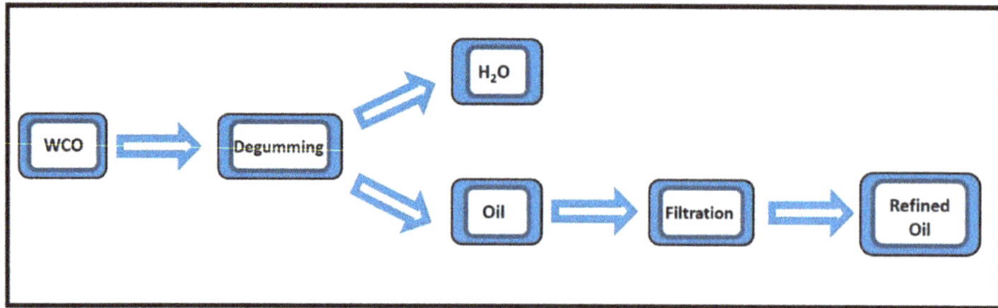

Figure 1. Process for WCO recycling, Mannu et al. [29].

The proposed process can be schematized as follows: WCO is subject to water treatment while the waste-water is discharged. The degummed waste oil is then pushed through a pressure pipeline until the filtering module, where it is treated with a porous material. Pressure filtration (4 bar) was described by the authors for pushing the intermediate oil through a fiberglass container containing the filtering material. A "deep bed" filtration was proposed, where the oil crosses the length of the filter. The combination between degumming and filtration affords a refined, recycled vegetable oil.

Despite the many progresses achieved in the field of such filters, further efforts are necessary to improve the performance and cost sustainability. In particular, fervent research is still required to decrease the amount of oil absorbed on the cellulose filters, to reduce the cost of synthesis and regeneration of synthetic materials and, last but not least, to understand the microscopic mechanisms behind the sequestration process, fundamental for improving the adsorbing properties of these materials up to their industrial scale-up and commercialization.

3.3. Separation Based on the Boiling Point (Distillation)

Distillation of WCOs also deserves to be described. In fact, the simplicity of such treatment, which is mainly aimed at removing volatile compounds and water from crude WCOs, allowed the commercial production of machines for small-scale WCO recycling. Nowadays, it is possible to buy a waste oil recycling machine for less than 15,000 USD [77].

These machines, for WCO recycling, are basically mini distillation plants. The general process applied to the raw material can be resumed as follows: the crude oil is subject to a gross filtration, aimed to remove solid contaminants, and then heated under vacuum. Through reduced pressure distillation, the crude oil is purified from the volatile fraction, resulting in a regenerate oleic product. After the deodorization procedure, the recycled vegetable oil can be sold as low quality lubricant, or as raw material for other industries, as fuel, high quality bio-lubricant, or soap productions.

This technology presents some evident advantages, as the reduced dimensions (from 1 m × 1.5 m × 1.2 m), fast processing time (flow rate from 10 L/min.), and low working temperature (less than 100 °C) [78]. On the other hand, it requires continuous electricity consumption, and it works under reduced pressure, which could reveal important issues in a working environment.

The energetic request is not very high (from 38 kW) and the process can be considered quite economic. It is possible to estimate a consumption of 1.583 kWh to produce 25 tons of purified oil, which corresponds to an overall energetic cost of approximately 7 € (euros) per ton of purified oil, based on the average energy prices in Europe for non-household consumers (0.11 €/kWh) [79].

Regarding the quality of the recycled oil, it depends on the specific configuration of the machine, as well as the quality of the raw WCO. As a matter of fact, these machines can be employed to produce low quality lubricants from waste, which can be employed or sold as such, or can be further processed for refinement. This technology can be used directly from the end-user, also, to regenerate exhausted lubricants employed in close circuits. In particular, the bio-lubricants market share is fast growing, and during previous years, showed a Compound Annual Growth Rate (CAGR) of 5%, with US and Europe playing the role of major consumers [80].

Recycled cooking oils obtained by distillation, filtering, or by combination of the formers, can be sold as chainsaw lubricants, or as solvents. In the first case, depending to the required quality, some further refining could be necessary, especially for increasing the oxidation stability and pour point, e.g., by esterification and/or by addition of antioxidants [81]. In the second case, recycled vegetable oil can be transformed in many derivatives, such as alkyds, poly(esteramide)s, poly(etheramide)s, polyurethanes, epoxies, and glycerol, which, blended with natural pigments, can find applications as biodegradable coatings and paints [82].

4. Conclusions

Waste cooking oils (WCOs) represent a valuable source of raw materials with a wide range of applications, from energy to lubricants or soaps. Moreover, due to their waste-nature and to a huge and diffused availability, WCOs fit well in the circular economy model, resulting in interest for their integration into existing processes, as well as for the development of new sustainable productions. Old techniques and technologies regarding the transformation of vegetable oils are, nowadays, object to specific updates in order to be adapted to the current characteristics requested by modern companies. This modernization in WCO treatment includes, not only a research activity on more convenient chemical syntheses, but also the development of new materials, and the definition of new combined treatments, which involve different techniques integrated. Looking at the recent bibliography related to the recycling of WCOs, a first chemical transformation of the crude material, mainly through esterification, represents a transversal topic; many authors deal with biodiesel as well as bio-lubricant applications. Nevertheless, technological advancements, both in terms of processes and materials, aimed to recycle WCOs in a more efficient and sustainable way, are gaining attention. In this regard, mini plants based on fast physical treatment of waste oils (mostly by distillation), are nowadays available on the market. Looking at the future, the integration between academic and industrial research activities will be fundamental in order to develop recycling treatments that are truly able to substitute existing productions, which involve non-waste raw materials.

Author Contributions: Conceptualization, A.M. (Alberto Mannu) and S.G.; methodology, A.M. (Alberto Mannu) and S.G.; investigation, J.I.P.; data curation, A.M. (Alberto Mannu) and S.G.; writing—original draft preparation, A.M. (Alberto Mannu) and S.G.; writing—review and editing, A.M. (Andrea Mele); visualization, S.G.; supervision, A.M. (Alberto Mannu); project administration, A.M. (Alberto Mannu) and S.G. All authors have read and agreed to the published version of the manuscript.

Funding: This research received no external funding.

Acknowledgments: The authors thank Sonia Martel Martin for the kind help in managing the collaboration with UBU and all of the people involved from the MSCA RISE 2019 project proposal WORLD (873005).

Conflicts of Interest: The authors declare no conflict of interest.

References

1. Marion, P.; Bernela, B.; Piccirilli, A.; Estrine, B.; Patouillard, N.; Guilbot, J.; Jérôme, F. Sustainable chemistry: How to produce better and more from less? *Green Chem.* **2017**, *19*, 4973–4989. [CrossRef]
2. Rybicka, J.; Tiwari, A.; Leeke, G.A. Technology readiness level assessment of composites recycling technologies. *J. Clean. Prod.* **2016**, *112*, 1001–1012. [CrossRef]
3. De Almeida, S.T.; Borsato, M. Assessing the efficiency of End of Life technology in waste treatment—A bibliometric literature review. *Resour. Conserv. Recycl.* **2019**, *140*, 189–208. [CrossRef]
4. Borrello, M.; Caracciolo, F.; Lombardi, A.; Pascucci, S.; Cembalo, L. Consumers' Perspective on Circular Economy Strategy for Reducing Food Waste. *Sustainability* **2017**, *9*, 141. [CrossRef]
5. European Waste Catalogue, Code Number 20 01 25. Available online: https://eur-lex.europa.eu/legal-content/EN/TXT/PDF/?uri=CELEX:02000D0532-20150601&from=EN (accessed on 10 February 2020).
6. Choe, E.; Min, D.B. Chemistry of Deep-Fat Frying Oils. *J. Food Sci.* **2007**, *72*, R77–R86. [CrossRef] [PubMed]

7. Statista-The portal for statistics. Consumption of Vegetable Oils Worldwide From 2013/14 To 2017/2018, by Oil Type (in Million Metric Tons). Available online: https://www.statista.com/statistics/263937/vegetable-oils-globalconsumption (accessed on 10 February 2020).
8. Lin, C.S.K.; Pfaltzgraff, L.A.; Herrero-Davila, L.; Mubofu, E.B.; Abderrahim, S.; Clark, J.H.; Koutinas, A.A.; Kopsahelis, N.; Stamatelatou, K.; Dickson, F.; et al. Food waste as a valuable resource for the production of chemicals, materials and fuels. Current situation and global perspective. *Energy Environ. Sci.* **2013**, *6*, 426–464. [CrossRef]
9. Gurbuz, I.B.; Ozkan, G. Consumers' knowledge, attitude and behavioural patterns towards the liquid wastes (cooking oil) in Istanbul, Turkey. *Environ. Sci. Pollut. Res.* **2019**, *26*, 16529–16536. [CrossRef]
10. Liu, T.; Liu, Y.; Luo, E.; Wu, Y.; Li, Y.; Wu, S. Who is the most effective stakeholder to incent in the waste cooking oil supply chain? A case study of Beijing, China. *Energy Ecol. Environ.* **2019**, *4*, 116–124. [CrossRef]
11. Tsai, W.-T. Mandatory Recycling of Waste Cooking Oil from Residential and Commercial Sectors in Taiwan. *Resources* **2019**, *8*, 38. [CrossRef]
12. Ribau-Teixeira, M.; Nogueira, R.; Nunes, L.M. Quantitative assessment of the valorisation of used cooking oils in 23 countries. *Waste Manag.* **2018**, *78*, 611–620. [CrossRef]
13. Rodrigues Pereira Ramos, T.; Gomes, M.I.; Barbosa-Póvoa, A.P. Planning waste cooking oil collection systems. *Waste Manag.* **2013**, *33*, 1691–1703. [CrossRef] [PubMed]
14. Karmakar, G.; Ghosh, P.; Sharma, B. Chemically Modifying Vegetable Oils to Prepare Green Lubricants. *Lubricants* **2017**, *5*, 44. [CrossRef]
15. Hazrat, M.A.; Rasul, M.G.; Khan MM, K.; Ashwath, N.; Rufford, T.E. Emission characteristics of waste tallow and waste cooking oil based ternary biodiesel fuels. *Energy Procedia* **2019**, *160*, 842–847. [CrossRef]
16. Ray, S.K.; Prakash, O. Biodiesel Extracted from Waste Vegetable Oil as an Alternative Fuel for Diesel Engine: Performance Evaluation of Kirlosker 5 kW Engine. In *Renewable Energy and its Innovative Technologies*; Springer: Berlin/Heidelberg, Germany, 2019.
17. Chrysikou, L.P.; Dagonikou, V.; Dimitriadis, A.; Bezergianni, S. Waste cooking oils exploitation targeting EU 2020 diesel fuel production: Environmental and economic benefits. *J. Clean. Prod.* **2019**, *219*, 566–575. [CrossRef]
18. Rayhan, B.A.; Kamal, H. Waste cooking oil as an asphalt rejuvenator: A state-of-the-art review. *Constr. Build. Mater.* **2020**, *230*, 116985.
19. Magrinyà, N.; Tres, A.; Codony, R.; Guardiola, F.; Nuchi, C.D.; Bou, R. Use of recovered frying oils in chicken and rabbit feeds: Effect on the fatty acid and tocol composition and on the oxidation levels of meat, liver and plasma. *Animal* **2012**, *7*, 505–517.
20. Chung, T.Y.; Eiserich, J.P.; Shibamoto, T. Volatile compounds identified in headspace samples of peanut oil heated under temperatures ranging from 50 to 200.degree.C. *J. Agric. Food Chem.* **1993**, *41*, 1467–1470. [CrossRef]
21. Wu, C.M.; Chen, S.Y. Volatile compounds in oils after deep frying or stir frying and subsequent storage. *J. Am. Oil Chem. Soc.* **1992**, *69*, 858–865. [CrossRef]
22. Mannu, A.; Ferro, M.; Di Pietro, M.E.; Mele, A. Innovative Applications of Waste Cooking Oil as Raw Material. *Sci. Prog.* **2019**, *102*, 153–160. [CrossRef]
23. Panadare, D.C.; Rathod, V.K. Applications of Waste Cooking Oil Other Than Biodiesel: A Review. *Iran. J. Chem. Eng.* **2015**, *12*, 55–76.
24. Singhabhandhu, A.; Tezuka, T. The waste-to-energy framework for integrated multi-waste utilization: Waste cooking oil, waste lubricating oil, and waste plastics. *Energy* **2010**, *35*, 2544–2551. [CrossRef]
25. Awogbemi, O.; Onuh, E.I.; Inambao, F.L. Comparative study of properties and fatty acid composition of some neat vegetable oils and waste cooking oils. *Int. J. Low Carbon Technol.* **2019**, *14*, 417–425. [CrossRef]
26. Ziaiifar, A.M.; Achir, N.; Courtois, F.; Trezzani, I.; Trystram, G. Review of mechanisms, conditions, and factors involved in the oil uptake phenomenon during the deep-fat frying process. *Int. J. Food Sci. Technol.* **2008**, *43*, 1410–1423. [CrossRef]
27. Saguy, I.; Dana, D. Integrated approach to deep fat frying: Engineering, nutrition, health and consumer aspects. *J. Food Eng.* **2003**, *56*, 143–152. [CrossRef]
28. Pokorny, J. Flavor chemistry of deep fat frying in oil. In *Flavor Chemistry of Lipid Foods*; Min, D.B., Smouse, T.H., Zhang, S.S., Eds.; American Oil Chemists Society: Champaign, IL, USA, 1989.

29. Mannu, A.; Ferro, M.; Colombo Dugoni, G.; Panzeri, W.; Petretto, G.L.; Urgeghe, P.; Mele, A. Recycling of Waste Cooking Oils: Variation of the Chemical Composition during Water Treatment. *Preprints* **2019**, *160*, 842–847. [CrossRef]
30. Li, J.; Cai, W.; Sun, D.; Liu, Y. A Quick Method for Determining Total Polar Compounds of Frying Oils Using Electric Conductivity. *Food Anal. Methods* **2016**, *9*, 1444–1450. [CrossRef]
31. Firestone, D. *Deep Frying: Chemistry, Nutrition, and Practical Applications*, 2nd ed.; Erickson, M.D., Ed.; Elsevier: Amsterdam, The Netherlands, 2007; pp. 373–385. [CrossRef]
32. Mannu, A.; Ferro, M.; Colombo Dugoni, G.; Panzeri, W.; Petretto, G.L.; Urgeghe, P.; Mele, A. Improving the recycling technology of waste cooking oils: Chemical fingerprint as tool for non-biodiesel application. *Waste Manag.* **2019**, *96*, 1–8. [CrossRef]
33. Fu, J.; Turn, S.Q.; Takushi, B.M.; Kawamata, C.L. Storage and oxidation stabilities of biodiesel derived from waste cooking oil. *Fuel* **2016**, *167*, 89–97. [CrossRef]
34. Reyero, I.; Arzamendi, G.; Zabala, S.; Gandía, L.M. Kinetics of the NaOH-catalyzed transesterification of sunflower oil with ethanol to produce biodiesel. *Fuel Process. Technol.* **2015**, *129*, 147–155. [CrossRef]
35. García Martín, J.F.; del Carmen López Barrera, M.; Torres García, M.; Zhang, Q.-A.; Álvarez Mateos, P. Determination of the Acidity of Waste Cooking Oils by Near Infrared Spectroscopy. *Processes* **2019**, *7*, 304. [CrossRef]
36. Banerjee, A.; Chakraborty, R. Parametric sensitivity in transesterification of waste cooking oil for biodiesel production—A Review. *Resour. Conserv. Recycl.* **2009**, *53*, 490–497. [CrossRef]
37. Félix, S.; Araújo, J.; Pires, A.M.; Sousa, A.C. Soap production: A green prospective. *Waste Manag.* **2017**, *66*, 190–195. [CrossRef]
38. Fereidooni, L.; Tahvildari, K.; Mehrpooya, M. Trans-esterification of waste cooking oil with methanol by electrolysis process using KOH. *Renew. Energy* **2018**, *116*, 183–193. [CrossRef]
39. Cai, Z.-Z.; Wang, Y.; Teng, Y.-L.; Chong, K.-M.; Wang, J.-W.; Zhang, J.-W.; Yang, D.-P. A two-step biodiesel production process from waste cooking oil via recycling crude glycerol esterification catalyzed by alkali catalyst. *Fuel Process. Technol.* **2015**, *137*, 186–193. [CrossRef]
40. Álvarez-Mateos, P.; García-Martín, J.F.; Guerrero-Vacas, F.J.; Naranjo-Calderón, C.; Barrios-Sánchez, C.C.; Pérez-Camino, M.C. Valorization of a high-acidity residual oil generated in the waste cooking oils recycling industries. *Grasas Aceites* **2019**, *70*, e335. [CrossRef]
41. Quitain, A.T.; Sumigawa, Y.; Mission, E.G.; Sasaki, M.; Assabumrungrat, S.; Kida, T. Graphene Oxide and Microwave Synergism for Efficient Esterification of Fatty Acids. *Energy Fuels* **2018**, *32*, 3599–3607. [CrossRef]
42. Costa, E.; Cruz, M.; Alvim-Ferraz, C.; Almeida, M.F.; Dias, J.M. WASTES—Solutions, Treatments and Opportunities II. In Proceedings of the 4th International Conference Wastes: Solutions, Treatments and Opportunities, WASTES 2017, Porto, Portugal, 25 September 2017.
43. Nurdin, S.; Yunus, R.M.; Nour, A.H.; Gimbun, J.; Aisyah, N.; Azman, N.; Sivaguru, M.V. Restoration of Waste Cooking Oil (WCO) Using Alkaline Hydrolysis Technique (ALHYT) for Future Biodetergent. *ARPN J. Eng. Appl. Sci.* **2016**, *11*, 6405–6410.
44. Liu, W.-Q.; Yang, S.-Z.; Gang, H.-Z.; Mu, B.-Z.; Liu, J.-F. Efficient emulsifying properties of monoglycerides synthesized via simple and green route. *J. Dispers. Sci. Technol.* **2019**, 1–9. [CrossRef]
45. Mićić, R.; Tomić, M.; Martinović, F.; Kiss, F.; Simikić, M.; Aleksic, A. Reduction of free fatty acids in waste oil for biodiesel production by glycerolysis: Investigation and optimization of process parameters. *Green Process. Synth.* **2019**, *8*, 15–23. [CrossRef]
46. Kombe, G.G. Re-esterification of high free fatty acid oils for biodiesel production. *Biofuels* **2015**, *6*, 31–36. [CrossRef]
47. Supardan, M.D.; Adisalamun Lubis, Y.M.; Annisa, Y.; Satriana Mustapha, W.A.W. Effect of co-solvent addition on glycerolysis of waste cooking oil. *Pertanika J. Sci. Technol.* **2017**, *25*, 1203–1210.
48. Mazubert, A.; Crockatt, M.; Poux, M.; Aubin, J.; Roelands, M. Reactor Comparison for the Esterification of Fatty Acids from Waste Cooking Oil. *Chem. Eng. Technol.* **2015**, *38*, 2161–2169. [CrossRef]
49. Farid, M.A.A.; Hassan, M.A.; Taufiq-Yap, Y.H.; Ibrahim, M.L.; Othman, M.R.; Ali AA, M.; Shirai, Y. Production of methyl esters from waste cooking oil using a heterogeneous biomass-based catalyst. *Renew. Energy* **2017**, *114*, 638–643. [CrossRef]
50. Borovinskaya, E.S.; Sabaditsch, D.; Reschetilowski, W. Base-Catalyzed Ethanolysis of Waste Cooking Oil in a Micro/Millireactor System: Flow and Reaction Analysis. *Chem. Eng. Technol.* **2019**, *42*, 495–505. [CrossRef]

51. Steven, J.R. Synthetics, mineral oils, and bio-based lubricants: Chemistry and technology. In *Chemical Industries*; Rudnick, R.L., Ed.; CRC Press: Boca Raton, FL, USA, 2006.
52. Yunus, R.; Fakhru'l-Razi, A.; Ooi, T.L.; Biak DR, A.; Iyuke, S.E. Kinetics of transesterification of palm-based methyl esters with trimethylolpropane. *J. Am. Oil Chem. Soc.* **2004**, *8*, 497–503. [CrossRef]
53. Sun, G.; Li, Y.; Cai, Z.; Teng, Y.; Wang, Y.; Reaney, M.J.T. K_2CO_3-loaded hydrotalcite: A promising heterogeneous solid base catalyst for biolubricant base oil production from waste cooking oils. *Appl. Catal. B Environ.* **2017**, *209*, 118–127. [CrossRef]
54. Costarrosa, L.; Leiva-Candia, D.E.; Cubero-Atienza, A.J.; Ruiz, J.J.; Dorado, M.P. Optimization of the Transesterification of Waste Cooking Oil with Mg-Al Hydrotalcite Using Response Surface Methodology. *Energies* **2018**, *11*, 302. [CrossRef]
55. Mannu, A.; Ferro, M.; Colombo Dugoni, G.; Garroni, S.; Taras, A.; Mele, A. Response Surface Analysis of density and flash point in recycled Waste Cooking Oils. *Chem. Data Collect.* **2020**, *25*, 100329. [CrossRef]
56. Gupta, M.K. *Practical Guide to Vegetable Oil Processing*, 2nd ed.; Academic Press and AOCS Press: New York, NY, USA, 2017; ISBN 9781630670504.
57. Vlahopoulou, G.; Petretto, G.L.; Garroni, S.; Piga, C.; Mannu, A. Variation of density and flash point in acid degummed waste cooking oil. *J. Food Process. Preserv.* **2018**, *42*, e13533. [CrossRef]
58. Pellati, F.; Benvenuti, S.; Yoshizaki, F.; Bertelli, D.; Rossi, M.C. Head Space Solid Phase Micro Extraction Gas Chromatography coupled with Mass. *J. Chromatogr. A* **2005**, *1087*, 265–273. [CrossRef]
59. Ladhe, A.R.; Kumar, N.S.K. Application of Membrane Technology in Vegetable Oil Processing. *Membr. Technol.* **2010**. [CrossRef]
60. Koris, A.; Vatai, G. Dry degumming of vegetable oils by membrane filtration. *Desalination* **2002**, *148*, 149–153. [CrossRef]
61. Ochoa, N.; Pagliero, C.; Marchese, J.; Mattea, M. Ultrafiltration of vegetable oils: Degumming by polymeric membranes. *Sep. Purif. Technol.* **2011**, *22–23*, 417–422. [CrossRef]
62. Pagliero, C.; Ochoa, N.; Marchese, J.; Mattea, M. Vegetable oil degumming with polyimide and polyvinylidene fluoride ultrafiltration membranes. *J. Chem. Technol. Biotechnol.* **2004**, *79*, 148–152. [CrossRef]
63. Desai, N.C.; Mehta, M.H.; Dave, A.M.; Mehta, J.N. Degumming of vegetable oil by membrane technology. *Indian J. Chem. Technol.* **2002**, *9*, 529–534.
64. De Moura, J.M.; Gonçalves, L.A.; Petrus, J.C.; Viotto, L.A. Degumming of vegetable oil by microporous membrane. *J. Food Eng.* **2005**, *70*, 473–478. [CrossRef]
65. Tur, E.; Onal-Ulusoy, B.; Akdogan, E.; Mutlu, M. Surface Modification of Polyethersulfone Membrane to Improve Its Hydrophobic Characteristics for Waste Frying Oil Filtration: Radio Frequency Plasma Treatment. *J. Appl. Plymer Sci.* **2011**, *123*, 3402–3411. [CrossRef]
66. Casallas, I.D.; Carvajal, E.; Mahech, E.; Castrillón, C.; Gómez, H.; López, C.; Malagón-Romero, D. Pre-treatment of Waste Cooking Oils for Biodiesel Production. *Chem. Eng. Trans.* **2018**, *65*, 385–390. [CrossRef]
67. Weisshaar, R. Quality control of used deep-frying oils. *Eur. J. Lipid Sci. Technol.* **2014**, *116*, 716–722. [CrossRef]
68. Muraldihara, H.S.; Jirjis, B.F.; Seymour, G.F. Process for Removing Vegetable Oil Waxes by Fast Cooling Vegetable Oil and Using a Porous Non-Metallic Inorganic Filter. Available online: https://patents.google.com/patent/US5482633A/en (accessed on 19 March 2020).
69. Bertram, B.; Abrams, C.; Kauffman, J. Adsorbents Filtration System for Treating Used Cooking Oil or Fat in Frying Operations. Available online: https://patents.google.com/patent/US6368648B1/en (accessed on 19 March 2020).
70. Bivens, T.; Calrk, J.G. Filtration and Filtration Method for Cooking Oil Used in Frying Process. Available online: https://patents.google.com/patent/US20070289927A1/en (accessed on 19 March 2020).
71. Bivens, T.; Calrk, J.G. Method for Filtering Cooking Oil Used in Frying Process. Available online: https://patents.google.com/patent/US8066889B2/en (accessed on 19 March 2020).
72. For EU See Waste Oil Directive 75/439/EEC. Available online: https://ec.europa.eu/environment/archives/waste/oil/questionnaire.htm (accessed on 19 March 2020).
73. Available online: https://fryoilsaver.com/fry-oil-management/ (accessed on 4 March 2020).
74. Vlasova, E.A.; Yakimov, S.A.; Naidenko, E.V.; Kudrik, E.V.; Makarov, S.V. Application of metal–organic frameworks for purification of vegetable oils. *Food Chem.* **2016**, *190*, 103–109. [CrossRef]

75. Mannu, A.; Vlahopoulou, G.; Sireus, V.; Petretto, G.L.; Mulas, G.; Garroni, S. Bentonite as a Refining Agent in Waste Cooking Oils Recycling: Flash Point, Density and Color Evaluation. *Nat. Prod. Commun.* **2018**, *13*, 613–616. [CrossRef]
76. Mannu, A.; Vlahopoulou, G.; Urgeghe, P.; Ferro, M.; Del Caro, A.; Taras, A.; Garroni, S.; Rourke, J.P.; Cabizza, R.; Petretto, G.L. Variation of the Chemical Composition of Waste Cooking Oils upon Bentonite Filtration. *Resources* **2019**, *8*, 108. [CrossRef]
77. Available online: https://www.lushunoilpurifier.com/product/productconnect?id=166&gclid=Cj0KCQiA04XxBRD5ARIsAGFygj_56-c-8FQiNadDYxYOHfUj4SXlaS0NAN5hLibRoBGB7bRxoHSOmi8aApDmEALw_wcB (accessed on 20 February 2020).
78. Available online: https://www.alibaba.com/product-detail/Stainless-Steel-Cooking-Oil-Filter-Machine_62090539203.html?spm=a2700.drainage_lp_1.0.0.43391199tvsgjr&s=p (accessed on 20 February 2020).
79. Available online: https://ec.europa.eu/eurostat/statistics-explained/index.php?title=Electricity_price_statistics/es#Precios_de_la_electricidad_para_los_consumidores_dom.C3.A9sticos (accessed on 20 February 2020).
80. Luzuriaga, S. Bio-lubricants: Global opportunities and challenges. *Tribol. Lubr. Technol.* **2017**, *73*, 16–19.
81. Kofi, S.; Kotoka, T.F. The potential of castor, palm kernel, and coconut oils as biolubricant base oilvia chemical modification and formulation. *Therm. Sci. Eng. Progress* **2020**, *16*, 1004802. [CrossRef]
82. Manawwer, A.; Deewan, A.; Eram, S.; Fahmina, Z.; Sharif, A. Vegetable oil based eco-friendly coating materials: A review article. *Arab. J. Chem.* **2014**, *7*, 469–479. [CrossRef]

MDPI
St. Alban-Anlage 66
4052 Basel
Switzerland
Tel. +41 61 683 77 34
Fax +41 61 302 89 18
www.mdpi.com

Processes Editorial Office
E-mail: processes@mdpi.com
www.mdpi.com/journal/processes